Human Peoples

Human Peoples

On the Genetic Traces of Human Evolution, Migration and Adaptation

LLUÍS QUINTANA-MURCI

Translated by Howard Curtis

ALLEN LANE

an imprint of

PENGUIN BOOKS

ALLEN LANE

UK | USA | Canada | Ireland | Australia
India | New Zealand | South Africa

Allen Lane is part of the Penguin Random House group of companies
whose addresses can be found at global.penguinrandomhouse.com

Penguin
Random House
UK

First published in Great Britain by Allen Lane 2024
001
Copyright © Lluís Quintana-Murci, 2024
Translation copyright © Howard Curtis, 2024

The moral rights of the author and translator have been asserted

Set in 12/14.75pt Dante MT Std
Typeset by Jouve (UK), Milton Keynes
Printed and bound in Great Britain by Clays Ltd, Elcograf S.p.A.

The authorized representative in the EEA is Penguin Random House Ireland,
Morrison Chambers, 32 Nassau Street, Dublin D02 YH68

A CIP catalogue record for this book is available from the British Library

ISBN: 978–0–241–60915–6

www.greenpenguin.co.uk

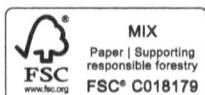

Contents

Introduction

Where do we come from? What are we?
Where are we going?

These are three questions that have haunted us over the centuries. For answers, we have looked to religion and philosophy, to art and history, as well as to science. These three questions form the title of Paul Gauguin's masterpiece from 1897–98, which now hangs in the Museum of Fine Arts in Boston. It was painted in Tahiti, where Gauguin had gone in search of values different from those of Western society, and needs to be read from right to left, in the opposite direction to that familiar to Europeans. On the right, to respond to the question 'Where do we come from?', there is a group of women with a baby; in the centre, the question 'What are we?' is illustrated by the daily lives of young adults and their relationship with nature; on the left, the question of the future, 'Where are we going?', receives an answer symbolizing old age and the afterlife.

The painting is an ode to travel, to difference, to human diversity – the diversity of individuals, of sexes, of generations, the diversity found in the places and times of our lives. As such, Gauguin's work resonates deeply with the image that population genetics gives us of the human world. The aim of this field is to study the genetic diversity of humanity – and to tell us about our past, about our ancestors' journeys. Over the last few decades, genetics has progressed to become an extremely powerful exploratory tool, a tool that, when applied to our species, reveals the extent of diversity among individuals and groups – the dazzling variety of the human fabric that covers our planet. It reveals the molecular secrets of organisms, shining a light on biological variation, those differences shaped by the constraints of geography and ecology, with a degree

I

of precision that opens up a new dimension in our knowledge of humanity. It is a tool that does not merely provide a snapshot of our present but allows us also to travel back in time and obtain an image of our species in motion, restoring in detail the way in which our shapes, sizes and physiologies were established over our long settlement of the Earth.

As we shall see throughout this book, the discoveries of population genetics may be used to unlock the secrets of our history, our evolution and our adaptations to changing environments. They also open up a new world for medicine, which can use this knowledge to prevent and treat our diseases.

This book is intended as an allegory of human diversity: we are a species composed of many peoples, enriched by our differences. The focus will be above all on genetic diversity, but not exclusively. To discover, analyse and interpret humanity in all its forms, population genetics requires other sources of difference, such as the geographical diversity of individuals, their linguistic affiliations, their lifestyles and modes of subsistence, as well as the whole range of their customs and socio-cultural organizations.

Today, genetics goes a long way beyond the bounds of biology. It has become an important way to write and understand history, because, to quote biologist Theodosius Dobzhansky's famous dictum, 'Nothing in biology makes sense except in the light of evolution.' To understand life in the present we need first to understand the process of evolution through time: creatures are what they have *become*. Knowing them also means tracing their history.

Humanity's history goes back to the most remote times, so remote in fact that only vestiges remain – for the most ancient humans, we have simply inherited a few fragments of bone. We leave, it may seem, very little behind us. 'Alas, poor Yorick!' Hamlet cries when he is handed a skull. But Yorick in fact provides much more than a tragic tribute to the finiteness of human existence: by exhuming the remains of the dead, we are gifted, across the many millennia that separate us from their lifetimes, something of the living creatures they once were. Thanks to genetics, these bones, and the DNA they contain, allow us to reconstruct a little of their history and that of the species to which they belong, and in this way claim them for ourselves, to reintegrate them into the great biological family that is the human species. Bones speak – at least to scientists. They are a fossil archive, which needs to be deciphered. We have learnt, over the centuries, to recognize the human traces buried underground and to date them: the planet has become a formidable spatial-temporal puzzle where we find many clues to the history of life and of our species. But they must still be put in order, and it is here that genetics contributes crucial new knowledge. In 1953, the discovery of the structure of DNA inaugurated the era of molecular biology and genetics, which would revolutionize our knowledge of living things (and how to keep things alive, too: without it, there would have been no messenger RNA vaccines to guard us against COVID-19). It would also, paradoxically, transform our understanding of the dead – and therefore of our history. In *Jurassic Park*, Michael Crichton may have created a far-fetched work of imagination, but one that at least had the merit of popularizing a fundamental scientific fact: the DNA of creatures that lived thousands (or even millions!) of years ago may be found and reconstructed.

There is more. The present century began, in 2001, with the sequencing of the human genome: its 3 billion letters, or nucleotides, carry all the biological information that makes us what we are. And so for population genetics, the transition from the twentieth to the twenty-first century was concurrent with a great movement from genetics (which studies a specific gene that has a specific function) to genomics (which involves studying the entire genome, all our DNA, composed of around 20,000 genes). We no longer perceived just the individual, but also the population of which they were a part. In deciphering the genome, we were allowed at once to compare across genomes: a qualitative leap in our knowledge of living things. From the *century of the gene* (the title of a book by Evelyn Fox Keller), we have moved to the *century of the genome*. Genomics encompasses the full extent of diversity: it aims to be inclusive, broad, plural. It is constantly surprising us, telling us that our species cannot be reduced to *one* genome but is made up of countless genomes. Even on an individual level, the genome of each of us is in fact a mosaic of the multiple genomes of our ancestors, bearing witness to a history that has lasted at least 200,000 years.

This, then, is the starting point for this book, an expedition into the new knowledge of humanity that genomics provides. Genomics allows us to sequence and analyse the billions of base pairs of an individual's genome, to compare this genome with the characteristics of contemporary populations all over the world, then to place it back in the context of a history – the history of individuals and populations and their migrations across the planet. This history encompasses the evolution of our species, and even of those long-dead species from which we descend, at least in part. All this was completely unimaginable only two decades ago – and we are still only at the beginning.

For example, by analysing a single bone fragment of a young girl who lived more than 50,000 years ago in southern Siberia, it is now possible to say, as Svante Pääbo's team did in 2018, that her mother was a Neanderthal and her father a Denisovan – two human lineages that are now extinct. For the physicist and astronomer Galileo, the 'great book of nature' was written in the language of mathematics. For present-day biologists, it is written in the language of DNA. As we decipher these pieces of bone, they yield up their treasures, telling us about the humans that they were, their life, their world, thereby regaining their place in human history.

By discovering where we come from, we are gaining a greater understanding of who we are. The image that is taking shape in front of our eyes, as revealed by genomics, is an image of human diversity. This is not an ideological slogan or a well-intentioned mantra, but a scientific observation. It is this diversity, and the biological mechanisms that produce it, the history that embodies it, as well as the science and medicine that may derive from it, that this book will explore.

If our origins enlighten us as to what our species is, the converse is also true: the question 'What are we?' enlightens us as to our origins. It is the very diversity of current human genomes which has shown us not only that the cradle of humanity is in Africa, confirming what we already knew from archeological evidence, but that there was much greater variety among our African ancestors than was once thought. In the 1980s, researchers showed, on the basis of mitochondrial DNA, inherited through the mother, and the Y chromosome, inherited through the father, that the history of our ancestors could indeed be traced to Africa. People began to talk about the discovery of 'the genetic Adam and Eve' of all humans, arousing a great deal of excitement in the scientific community. In fact, in the 2000s, genomics showed that we have multiple genetic ancestors. They are all from Africa, but we can only find the traces of those whose genetic features have survived to the present day, and the trail of the female line and that of the male line can be quite distinct. It is therefore quite possible that those maternal ancestors whose genetic features are still present in our genomes, the supposed 'Eves', lived in East Africa, while our paternal ancestors, the 'Adams', may have lived, for example, in South Africa.

Studying the genomes of today's human populations also shows that some of our ancestors left Africa about 60,000 years ago to populate the rest of the planet. This was the first great migration in human history: all present-day individuals of non-African ancestry are the descendants of these first African 'migrants'. But this initial exodus from Africa was merely the beginning of a long odyssey of migrations that led to the settlement of Europe, Asia and Australia about 50,000 years ago, then the Americas less than 30,000 years ago, and much later the islands of Remote Oceania, which were peopled for the first time only 1,000 years ago. Beyond these large migrations across continents, modern population genetics has allowed us to go further and to discover and date other migratory events and demographic processes at an unprecedented level of resolution. It is also thanks to these studies that we know today that admixture (the mixing between different populations) has been a constant process throughout human history and that we are all, to different extents, admixed, since our genomes are made up of multitudes of segments of DNA from varied sources. Our species is a historical and geographical patchwork, in which peoples and generations mix.

There have been even more surprises associated with the study of our genomes. So-called 'modern' humans – in other words, us, Sapiens – and Neanderthals were for a long time considered to be different species who, by definition, had never admixed and had not had any descendants. The twenty-first century has shown the opposite: Sapiens and Neanderthals did admix and do have descendants in common. In fact, we all carry today in our genomes – at least those of us who are not of African origin – between 1% and 3% of genetic material derived from the Neanderthals. And the surprises do not stop there. Thanks to genetics, and to genetics alone, a finger bone found in a cave in Siberia, which was thought to be of Neanderthal origin, was identified, by means of DNA sequencing, as belonging to another form of ancient human: the Denisovans. Better still: the ancestors of today's Asians admixed with the Denisovans, and some present-day populations, for example those living in Papua New Guinea, carry in their genomes a Denisovan heritage estimated as between 3.5% and 5%. Not only did humans admix between populations of *Homo sapiens*, but they also admixed with other forms of archaic humans whose genes survive in us.

This diversity had beneficial effects. *Homo sapiens* is an omnipresent and colonizing species. Men and women are present everywhere on the planet: from the hot dry savannahs to the Far North with its cold climates and meagre sunlight, from humid tropical rainforests to the extreme and inhospitable conditions of life at high altitude, where there is little oxygen, such as the Himalayas or the Andes. Over the last twenty years, the genomes of populations exposed to such environmental conditions have taught us a great deal about the way humans adapt genetically to the climate, to nutritional resources and to pathogens, among other things. In particular – and this is a major discovery – it has shown us that admixture was an essential factor that made it possible for our ancestors to better adapt to the new environments they encountered during their journey across the planet. For example, it is from admixture with the Denisovans that the Tibetans acquired their genetic ability to live at high altitudes in extreme conditions deficient in oxygen; and from admixture with the Neanderthals that the first Europeans improved their ability to withstand cold and survive certain pathogens, particularly viruses.

Pathogens are especially important here, since they have been with us ever since we appeared on Earth. Their presence was probably the primary cause of death for our species until the improvement in hygienic conditions and the discovery, at the end of the nineteenth century and the beginning of the twentieth century, of the first vaccines and antibiotics. But the decrease in the rate of death caused by infectious diseases is visible only in those countries that have access to modern medicine, which unfortunately is not the case for many. Even in industrialized countries, we have seen the havoc a simple pathogen can wreak, most recently with the SARS-CoV-2 coronavirus. This is a startling reminder of the vulnerability of our species to sudden and unpredictable changes of environment.

Try to imagine the COVID-19 crisis without hospitals, without respirators, without antibiotics, without hygiene and without any possibility of developing a vaccine. These are the very conditions we endured for more than 99% of our history. The effects in terms of human mortality were such that we can detect its imprint in our genomes today. And it is thanks to the study of the imprint of natural selection exerted by pathogens in the past that we can now identify the human genes that have played, and still play, a key role in winning the perpetual arms race against infectious diseases.

Nevertheless, because of the changing nature of the environment over time, there may be collateral damage in the adaptation of humans to their environment. In fact, what may have led to an *adaptation* in the past may be transformed, after a change in environment or lifestyle, into a *maladaptation* and lead to the emergence of certain present-day conditions such as auto-immune diseases, allergies, hypertension and obesity.

The study of the evolution of our genes – particularly those associated with the immune response – has proved to be a remarkably fertile area, complementing immunology, clinical genetics and epidemiology. It has improved our understanding of the factors, both genetic and non-genetic, that are associated with our different ways of coping with infections. With the progress achieved in the last twenty years in human genomics and in new methods of analysing big data, the adage that 'Nothing in biology makes sense except in the light of evolution' returns to us more meaningful than ever. And, to the extent that our knowledge of genetics has become a major help in the improvement of therapeutic effectiveness, we can go even further and state that 'Nothing in *medicine* makes sense except in the light of evolution.'

We are the product of our past, of our odysseys across the globe, of our adaptation to the environment and our constant admixture, both with human forms now extinct and with other populations of *Homo sapiens*. To get to know this history, this mix of natural evolution and human action that has invented new ways of life and so transformed our environment, genomics has developed methods that allow us to examine data with an unprecedented sharpness and accuracy. This knowledge is all the more valuable in that it contains a promise: that it may give us the keys to develop future medicines better adapted to each individual, known as 'precision medicine'. By getting to know individuals in all their genetic detail, we will be able to treat them more appropriately. By getting to know nature and its mechanisms better, we will have one more tool to remedy our weaknesses and fight more effectively against infectious diseases. As Louis Pasteur, the father of microbiology, said: 'Nature is the best doctor, she cures three-quarters of illnesses and never speaks ill of her colleagues.'

From Darwin to Genomics

Where do we come from? Mythologies and religions have proposed count-less origin stories for Earth and for life – many of which remain stuck in our collective imagination and continue to leave durable marks on our cultures. Science provides us with different answers; one of its aims is to describe and explain what we can glean of our origins from the processes of life that we perceive all around us. The methods to do so have grown in number over time and are still being perfected, but the general framework is provided by Darwin's theory of evolution. And that will be our starting point. To it, we add the discovery of DNA and the foundations of genetics, followed by population genetics, which led to our current knowledge of the diversity of our genomes, a remarkable instrument for knowing today's humans, with all that they owe to their past.

As far as population genetics is concerned, it all began in 1859, with the publication of Darwin's masterpiece *On the Origin of Species*, which marked the beginning of the evolutionary era. Barely a century and a half later, with the sequencing of the human genome in 2001, the genomic revolution was launched. Between these two dates, a series of theoretical and technological discoveries and developments gave us a greater understanding of how the human species evolved and suggested new answers to the question 'What are we?'

Population genetics rests on two disciplines: evolution and genetics. These two sciences were founded separately, in the mid-nineteenth century, by Charles Darwin and Gregor Mendel respectively. But it would be many decades before biologists made a connection between the fundamental processes of evolution and the principles of heredity. Of course, the idea of evolution had had its share of precursors and intuitions. In ancient times, thinkers like Anaximander of Miletus or Empedocles made timid allusions to the notion of evolution, of change, and the possibility that the origin of life was not, in fact, supernatural. Later, during the Enlightenment, there were dazzling insights, like those of Diderot, who, particularly in the *Encyclopédie*, emphasized the importance of the study of biology, the natural world and how species might change over generations. But the lasting influence of depictions supporting the idea that living creatures had a divine origin, and in particular the weight of 2,000 years of Christianity, prevented the emergence of evolutionary thinking before the beginning of the nineteenth century.

Once this obstacle had been removed, the seeds planted by Darwin quickly germinated. And the last century produced a rich harvest, giving us the principles of population genetics and the discovery of the foundations of heredity – DNA – as well as beginning to offer us an understanding of the diversity of our species, its migrations and its adaptation to its environment. Last but not least, the sequencing of fossil DNA revealed that in the course of its history our species exchanged genetic material with other human forms that are now extinct, like the Neanderthals or their Asian counterparts, the Denisovans. And this is only the beginning.

'Natura non facit saltus': Darwin's gradual evolution

Other thinkers had already raised the question of evolution. Jean-Baptiste de Lamarck (in his book *Philosophie zoologique*), Georges-Louis Leclerc de Buffon, Benoît de Maillet and Erasmus Darwin (Charles's grandfather) had questioned how static species were and raised the possibility that they might instead be subject to change. But it is Charles Darwin who is credited with introducing the theory of evolution as we know it today, even though he himself avoided the word 'evolution' in his book, preferring the expression 'descent with modification'.

Darwin's theory of evolution was based on a large number of observations, mostly made during his five-year voyage on board a survey ship, the *Beagle*, between 1831 and 1836, an event he himself considered the most important of his life. Under the command of Captain Robert FitzRoy, who had hired him as a young volunteer naturalist for a map-making mission around the world, Darwin collected and observed specimens wherever he went: from the Cabo Verde Islands to Amazonia, from Tierra del Fuego to the Galapagos and the Pacific. Once back in England, first in Cambridge and then in Kent, it would take him thirteen years to transform all his travel notebooks into a manuscript which he would finally publish in November 1859, for fear that Alfred Russel Wallace, who was about to publish a rather similar theory, might claim to have formulated these ideas first. Darwin's book was one of the few that have changed our view of the world. Its full title was *On the Origin of Species by Means of Natural Selection.*

The main premise of Darwin's theory is that species are transformed through natural selection: a gradual transformation, in accordance with the gradients observed in nature, hence Gottfried Wilhelm Leibniz's axiom *Natura non facit saltus* ('nature does not make jumps'), or the law of continuity. Darwin was the first to say that all individuals living on Earth have a common ancestor, and that the differences characterizing each of the species populating the planet were acquired through natural selection, which he took to be the fundamental mechanism of change, adaptation to the environment and subsequent speciation.

Charles Darwin championed the idea that finite resources such as food are the principal limitation to the growth of a population. This meant that there was competition between individuals or species for these resources, limiting their ability to survive and reproduce. In addition, Darwin postulated that the differences observed between individuals or species are 'transmitted' to the following generations, although he was not familiar with the concept of *genetic* transmission, and the underlying mechanism remained vague in Darwin's work. It is these differences that affect the ability of individuals or species to survive and reproduce – what we now call *fitness*. Over time, natural selection brings about gradual changes in the population, and the best-adapted individuals will be increasingly numerous.

This theory, which was revolutionary and very controversial at the time, marked the beginning of evolutionary thinking, even though, as we have seen, his contemporary Alfred Russel Wallace had reached the same conclusions as him. The two scientists even collaborated on an article about the theory of natural selection, which appeared in 1858. After the publication of his book, Charles Darwin led a solitary life in Kent, hurt by the reactions and the controversies it had stirred. He died in 1882, at the age of seventy-three.

Darwin's contemporary, the Austrian abbot Gregor Mendel, is the founding father of genetics – the second discipline that forms the basis of population genetics. Thanks to his work on the transmission of inherited traits published in 1866, heredity, which Darwin had hinted at, became a scientific concept describing an observable and manipulable reality. Mendel conducted thousands of experiments on 30,000 pea plants with different characteristics of colour, texture, and so on, crossing them and observing the way their characteristics were distributed in their descendants. Mendel showed that 'factors', which at the time were not yet called genes, were passed down from generation to generation in a predictable manner, thus establishing three laws of heredity, which would later be known as 'Mendel's laws'. Neglected at first, these laws would be rediscovered in 1900 by Hugo de Vries, Carl Correns and Erich von Tschermak. They seemed to apply exclusively to discrete, rather than continuous, characteristics, which meant that they went against Darwin's continuity theory. It is also to Hugo de Vries that we owe the term 'pangene', to describe the physical unit of transmission of characteristics, and to the Danish botanist and geneticist Wilhelm Johannsen the terms 'genetic' and 'gene'.

Unlike Darwin's work, which had an immense impact, Mendel's was not appreciated by his contemporaries. For three decades, it was barely noticed, and Charles Darwin never read the results. The link between heredity and evolution had to wait until the twentieth century to be established.

The birth of population genetics

The reconciliation between Darwinism and Mendelism began between the two world wars, with three young researchers from Francis Galton's British biometric school. These scientists, Ronald Fisher (1890–1962), Sewall Wright (1889–1988) and J. B. S. Haldane (1892–1964), founded population genetics, a discipline through which evolutionary biology and genetics formed a coherent, mathematically modelled whole.

Ronald Fisher laid the foundations of quantitative genetics, a fast-growing discipline today, where the genetic basis of complex diseases is being studied. Quantitative genetics deals with the statistical study of continuous phenotypes, such as height. Its hypothesis is that many genes contribute to the variability of the phenotype. One of Ronald Fisher's most important contributions was to show that the variability of a continuous phenotype is compatible with Mendelian inheritance. In his book *The Genetical Theory of Natural Selection*, he developed his basic theorem that evolution works through the natural selection of genetic mutations. He suggested that those mutations which have a strong impact on the phenotype have a greater chance of decreasing the fitness of individuals, while mutations with a weak effect have a higher chance of improving it and thus being favoured by natural selection. The evolution and transformation of phenotypes are therefore done gradually through the action of several mutations that have a weak effect – as Darwin had predicted. *Natura non facit saltus!*

Sewall Wright is best known for two concepts: *genetic drift* and *fitness landscape*. Genetic drift describes the random fluctuation, over the generations, in the frequency of mutations within a population. It is one of the major forces that shape genetic diversity. As for fitness landscape, this is a tool used in evolutionary biology to visualize the relationship between genetic mutations and the reproductive success of a population or a species, in other words, a depiction of fitness in the form of a map. Organisms may move across this adaptive landscape and climb to 'peaks' thanks to the acquisition of mutations that make them better adapted to their environment. Take the example of resistance to malaria. Genetic drift may lead an already adapted species either to the top of a peak – where it will be relatively resistant to malaria – or to the bottom of a valley – which means that it will be maladapted and vulnerable to malarial infection. This landscape is not definitive: a population that has 'come down' from a peak of fitness and is now in a 'valley', with a weak adaptive value, might then head towards the top of a higher peak than the previous one. In fact, if it benefits from new favourable mutations, under the effect of natural selection, it may become even more resistant to malaria than it was before.

J. B. S. Haldane is the third of the founding fathers of population genetics. He developed a mathematical approach to understanding how natural selection affects the frequency of mutations and how selection, mutation and migration interact. He is also known for advancing the hypothesis of a link between natural selection and resistance to malaria. In fact, the true father of this 1949 observation is the Italian geneticist Giuseppe Montalenti, who noted that disorders in the red blood cells, like the thalassemias or sickle-cell disease, were mainly observed in regions where malaria was endemic. It was not until 1954 that Anthony Allison confirmed the hypothesis, which is now considered a representative example of natural selection: disorders of the red blood cells may protect against malaria, which explains their increased frequency in regions where malaria is very common.

Discovery of the hereditary material: DNA

Although Ronald Fisher, Sewall Wright and J. B. S. Haldane laid the foundations for population genetics, we only see interdisciplinary consensus much later, between the 1930s and 1960s, with the collaboration of naturalists, paleontologists, mathematicians and geneticists that is now known as the 'modern synthesis'. Often described as neo-Darwinist, the modern synthesis marked the culmination of Darwin's theories. Its three most important proponents were Ernst Mayr (1904–2005), Theodosius Dobzhansky (1900–1975) and Julian Huxley (1887–1975). Their work proposed that evolution is a gradual process, as Darwin had suggested, compatible both with the known genetic mechanisms and with the observations of naturalists. The variability between individuals within a population is generated by mutation, recombination and gene flow. Evolution results from the combination of two mechanisms: on the one hand, the appearance within a population of new mutations and, on the other hand, natural selection or genetic drift acting on these mutations, changing their frequency in the population. Above all, this theory champions the idea that natural selection is the dominant force in evolution: it acts on the phenotypes of individuals in accordance with existing environmental conditions, leading to changes in the frequency of mutations that affect these phenotypes.

But, even though theoretical knowledge in population genetics was constantly increasing, the empirical data remained sparse, and the nature of the hereditary material was still unknown. The discovery of the double-helix structure of DNA in 1953 changed the future of population genetics for ever. Francis Crick (1916–2004), Rosalind Franklin (1920–1958) and James Watson (1928–) showed that DNA is composed of sequences of four different nucleotides, each consisting of a sugar linked to a phosphate group and a base containing nitrogen, notated as A, G, C or T. These nucleotides are arranged in a double helix of which the sugars and the phosphates constitute the skeleton. The bases point inwards and each of them is connected to a complementary base on the other strand by hydrogenous links.

It was thanks to these discoveries in molecular biology, and in genetics more generally, that the Japanese biologist Motoo Kimura (1924–1994) was able to combine theoretical approaches with empirical data, developing the neutral theory of molecular evolution in 1968. This theory postulates that most evolutionary changes are due to genetic drift, unlike the modern synthesis, which places the emphasis on the action of natural selection. The neutral theory does, however, refer to evolution at the molecular level, and Kimura himself acknowledged that phenotypic evolution cannot occur without the action of natural selection.

Population genetics has continued to provide us with new discoveries, often coupled with technological or methodological developments. Over the last fifty years, other noteworthy events have occurred, some of which we will explore later.

Exploring the sources of genetic diversity

How do the theoretical and mathematical models of population genetics help us to answer the question 'What are we?' Population genetics exists because we have become capable of understanding the effects, at the *macroscopic* level of populations over the course of human history, of *microscopic* changes at the level of genes and the molecular objects that underlie them. It is the theory of evolution that made this extraordinary synthesis possible, linking as it does all the levels of the phenomena of life in one unified explanation.

Indeed, the modern synthesis allows for a better understanding of the evolutionary processes responsible for changes in the frequency of mutations, over a given time and within a given space, in a given population. By understanding their mechanisms, we can produce mathematical models that approximate reality. Deciphering theoretically the interactions between processes such as natural selection and genetic drift allows us to deduce, from current genetic data, how these processes have shaped the genetic diversity of a population. Once we have adequate models, we still need to provide good data. In other words, if we can accurately model the past based on *theoretical* data, we become capable of deducing the past of a population based on *real* genetic data of today. That is the basic principle of studies in population genetics.

What are the evolutionary processes that affect the genetic diversity of a population? Decades of theoretical research in population genetics during the twentieth century showed that these evolutionary mechanisms may be divided into three broad categories: *genomic factors*, such as mutation or recombination, *demographic factors*, such as genetic drift or migration, and *selective factors*, which include the different forms that natural selection can take.

Genomic factors

Mutation is the only process that 'creates' diversity by producing molecular changes in DNA and generating new alleles. An allele is a variant of a given gene: each gene may have several 'versions' that produce different effects, as in the case of eye colour, for example. It could be said that mutation is the raw material of evolution, the process through which the evolutionary mechanisms are able to act. When a cell divides, it must replicate the DNA carrying its genome so that the two daughter cells will inherit the same genetic information as that contained within the mother cell. The double helix of DNA provides the perfect structure for simple replication: the two strands are unrolled to be separated, and each of the two strands serves as a model for recreating a strand in the complementary sequence by pairing between nucleic bases, which makes it possible to reconstruct two identical helices of duplex DNA. However, during the replication of the DNA, there may be 'errors', the original base (or nucleotide) being replaced by another one. Even though most of these errors are later corrected by a mechanism of proofreading and repair, they sometimes slip through the net and so give rise to mutations.

We can distinguish two large classes of mutations depending on the type of cells affected. On the one hand, there are the so-called *somatic* mutations: they do not affect the cells intended for reproduction and are therefore never transmitted. These mutations may appear throughout an individual's life in the DNA of any cell. In some cases, these may go on to become tumour cells. On the other hand, when the mutations affect the DNA of the stem cells of a gamete (spermatozoa and oocytes), they are called *germinal* mutations. In this case, the embryo will carry the mutation even though neither of the parents had it in their own genomes. In the case of germinal mutations, it seems that about 80% of the mutations passed on come from the material contributed by the father (the spermatozoon) and that the proportion of mutated spermatozoa is correlated with the age of the father. Nevertheless, anomalies contributed by the mother remain frequent and tend to increase with age too, but to a lesser extent. Today we know that on average each of us is born with 70 new mutations, compared with our parents, 55 inherited from the father and 15 from the mother.

Another source of genetic diversity comes from *recombination*, a process which, even though it does not create new genetic variants, does produce new genetic combinations, and therefore new genomes. In eukaryotes, that is organisms whose cells have a nucleus, recombination occurs during sexual reproduction and meiosis, the process of cell division of germ cells that produces the gametes such as the sperm or egg cells. The formation of new genetic combinations ensures genetic admixture and the maintenance of genetic diversity in a population, which increases the chances of a species adapting to environmental changes. Thus, mutation and recombination are evolutionary processes that increase both the genetic diversity within a population as well as the differences we find between populations.

Demographic factors

Another factor affecting the genetic diversity of populations is demography. First, it is worth mentioning *genetic drift* as defined by Sewall Wright. This is the modification of the frequency of an allele within a population, independently of mutation, natural selection and migration. Genetic drift is caused by purely random and unpredictable events associated with the encounters between spermatozoa and ova in the case of sexual reproduction. The extent of genetic drift is linked to the 'effective population size', which represents the number of individuals in a given population whose genetic diversity contributes to the following generation. In large populations, the frequency of mutations will remain relatively stable over many generations, since the effects of genetic drift are often negligible. In small populations, on the other hand, the impact of drift will be very marked, with major fluctuations of allelic frequencies over time: an advantageous allele may disappear, or a harmful allele may reach high frequency in a population.

There are some demographic events that are associated with a decrease in the size of the population and with more severe effects of drift. This is particularly the case with the geographical or cultural isolation of a population: the effects of drift can increase either through *bottlenecks*, characterized by a severe reduction in the size of a population following an environmental change, a war or an epidemic, for example, or through *founder effects*, that is, the establishment of a new population by a very small number of individuals belonging to a larger population. In these conditions, genetic drift tends to diminish the genetic diversity within the population, while it will exacerbate the differences between populations.

Migration, or gene flow, is another demographic mechanism that makes genetic exchanges between populations possible. The arrival of migrants may modify the distribution of the genetic diversity of the receiving population by modifying the frequency of its mutations. Migration and admixture prevent the development, between groups of populations, of major differences in their genetic make-up that might otherwise lead to complete speciation. In the case of a population with low genetic diversity following a founder effect or a bottleneck, the arrival of migrants and the ensuing admixture make it possible to re-establish the levels of genetic diversity, in a much quicker fashion than mutation. In this way, migration and mutation are forces antagonistic to genetic drift. Drift erodes diversity, but migrations and mutations re-establish it: this phenomenon is called the *mutation–drift balance*. Migration increases the genetic diversity within a given population, whereas it reduces the genetic differentiation between populations.

Selective factors

Finally, there are the factors that modify the patterns of genetic diversity of a population depending on the way in which *natural selection* manifests itself. Natural selection is the basis of biological adaptation to the environment. It occurs when there are differences in fitness between individuals. Since phenotypic characteristics are due partly to genetic variations, improving the chances of survival and reproduction of individuals, they are transmitted in the course of the generations. The influence of natural selection on genetic variability is complex, depending as it does on the type of selection put in place. It may increase, decrease, or maintain the levels of genetic diversity.

In the case of positive selection, an advantageous mutation will quickly become frequent in the population and improve fitness. By spreading widely and consequently reducing the frequency of other variants, this advantageous mutation leads to a decrease in genetic diversity within the population, while increasing the differences between human populations – assuming that the environmental pressure leading to such an event is different between the populations. But things do not stop there: there are other forms of natural selection – like 'balancing selection', which rests on the selective advantage of increased genetic diversity – that may increase the diversity within a population and reduce the differences between populations. A later chapter will examine the different forms of natural selection and how these affect the diversity of our genomes and our ability to adapt to the environment.

Reconstructing the past

The theoretical models developed in the second half of the twentieth century have made it possible for us to understand the relative contribution of each of these different factors to the genetic diversity of populations, and to establish estimates for the different demographic scenarios that characterize present-day populations. In doing this, researchers in population genetics have developed a body of knowledge that, combined with the abundance of population genomic data available today, is allowing them to also undertake the reverse process: no longer do they only explain the present based on the evolutionary past, but they can now start from the present state of populations to reconstruct their past. In fact, we can indeed recreate the human past based on current genomic data and estimate the important parameters of our evolution, such as the rate of growth of a certain population, the age of a mutation, or the rate of migration between separate groups.

A three-billion-dollar story: the Human Genome Project

Studies in genetics developed dramatically with the arrival of DNA sequencing techniques (that is, the techniques that allow us to determine the exact order of the nucleotides in the genome) in 1977, but progressed above all thanks to the Human Genome Project, a collaborative science programme launched in 1988 whose mission was to establish the complete DNA sequence of the human genome.

The first version of the genome corresponded to a 90% complete sequence that was finished in 2001, followed by an almost final version published in 2004. The difference between the first 'rough draft' and the final version is defined by the coverage of the DNA sequence, that is, the number of unsequenced gaps and the rate of error. In the 2004 version, there are fewer than 400 gaps, and 99% of the genome is determined with an accuracy rate of less than one error for every 10,000 base pairs. This work is still going on and the reference sequence of the genome is being constantly improved: in 2020, it was estimated that the number of unsequenced gaps is less than 100.

The Human Genome Project launched the era of modern human genetics, making it possible for the scientific community to use this sequence to carry out research both in medical genetics and in population genetics. The cost of the operation was astronomical: approximately 3 billion dollars (2.7 billion to be exact) was spent on sequencing the approximately 3 billion nucleotides (3.2 billion to be exact) that make up our genome. In 2021, twenty years after the first results arrived, techniques had progressed to such an extent that we can now sequence an entire genome in high resolution for less than 500 dollars. Still, whatever the price, this sequencing has made it possible to access key information on the structure as well as the content of the genome. We not only know today that the human genome is composed of about 3.2 billion nucleotides, we also know that only 2% of the genome is 'coding', that is, it contains genetic information that will translate into proteins, essential units for making a living being; that it harbours about 20,000 genes coding for proteins; and finally that the rest of the genome, some 98%, is non-coding and largely involved in regulatory functions.

The sequence of our genome comprises what is called nuclear DNA, divided throughout the 23 pairs of chromosomes located in the nucleus of each cell. Nevertheless, we also possess another small circular genome: mitochondrial DNA. This is found, as its name indicates, in the mitochondria – small organelles in the cell that produce the energy necessary to sustain it. In volume, it cannot be compared with nuclear DNA, containing as it does only 16,569 nucleotides and 37 genes – that is why it was the first genome to be sequenced, back in 1981. Nevertheless, as we shall see, despite its small size, it has delivered key information on the history of our species.

Each individual is unique

As far as the extent of genetic diversity between individuals or populations is concerned, the sequencing of the human genome initially gave very little information. It was in fact a single sequence: but each individual, apart from identical twins, has a genome unique to him or her. It was only with the arrival of new genomic technologies, such as microarrays and new-generation sequencers, combined with a decrease in the costs of sequencing, that the study of the *diversity* of the human genome entered its golden age. Several international consortiums were put together, like the HapMap Project or the 1000 Genomes Project, with the aim of characterizing human genetic diversity as a whole, and so facilitating studies in population genetics and medical genetics.

Thanks to them, we know today that the differences between the genomes of individuals rest on several categories of genetic variants. These variants are distinguished by their size – which may go from a single nucleotide to several hundreds of thousands – and their rate of mutation. For the variants that have been most studied and are most widespread in our genomes, the changes concern a single nucleotide. A well-known example is that of *single nucleotide polymorphisms* (SNPs), characterized by the replacement of one nucleotide by another in the DNA sequence (for example, C replaced by T). If we compare the genome of each of us to the 'reference sequence' (a representative sequence of a human genome used for comparative purposes), or if we compare two individuals chosen at random in the population, we will find on average about 3–4 million differences of the SNP type. In the vast majority of cases, these changes have no functional effect: either they are located outside the regions encompassing a gene, that is, regions that do not ultimately give rise to a protein, in which case they are called non-coding SNPs, or they are located in the exons of a gene (i.e. the coding part of the gene that determines a protein and therefore has a specific function, for example, darker skin pigmentation) but they do not lead to the replacement of an amino acid, in which case the SNPs are called synonymous. Most SNPs are neutral, in the sense that they do not affect the fitness of individuals. These neutral SNPs have proved extremely informative for the study of human population over time with regards to migration, admixture and fluctuations in population size.

In some cases, though, SNPs may modify the corresponding protein and ultimately lead to a phenotypic change, whether benign, like skin pigmentation, or associated with a disease, like cystic fibrosis. They are then called missense (or non-synonymous) SNPs, when they change the sequence of amino acids in the corresponding protein; or nonsense SNPs, when they lead to the replacement of a codon coding for an amino acid by a stop codon, generating a broken protein, generally non-functional.

The fact that an SNP is not located in a coding region does not mean that it has no functional effect. In fact, variants located in regulatory regions, like enhancers or promoters (i.e. fragments of DNA that control gene expression), may modify gene expression and thus alter the final phenotype.

But the variability we observe between the genomes of different individuals is not limited to SNPs. Each of us also differs from the reference sequence by about 450,000 small insertions or deletions ('indels'), that is, segments of DNA of fewer than 50 nucleotides that have been inserted or deleted from the DNA sequence. There is in addition a particular form of genetic variability whose importance has appeared more and more clearly over the last few years: 'structural' variations of the genome. They include deletions or duplications: they are called *copy number variations* (CNVs). They are different depending on the individual and can affect the number of copies of a particular gene or of a chromosomal segment in the genome.

CNVs provide a vital contribution to the diversity of the human genome and are therefore of great relevance in the framework of medical genetics and population genetics. Recent studies have shown that they may not only affect the dosage of genes, but also modulate the mechanisms regulating gene expression by altering the number of copies of the regulatory elements or modifying the three-dimensional structure of the genome. Other forms of structural variations exist in the genome, but to a lesser extent, such as large insertions, genomic rearrangements, or insertions of mobile elements.

A final form of variability involves sequences of the genome repeated in tandem. These are known as microsatellites if the size of the single motif is between 1 and 6 nucleotides, or minisatellites if it is between 7 and 49 nucleotides. Microsatellites are especially abundant in the human genome: there are several hundreds of thousands of them. Thanks to their high rate of mutation, they are used above all to compare human populations that are very close, or in the context of legal medicine. Microsatellites located in coding regions have been associated with more than 40 monogenic diseases, like the neurodegenerative diseases – such as Huntington's or Kennedy's – whereas those located in non-coding regions may also affect levels of gene expression.

Today, thanks to all these studies in genomics and to international consortiums, we know that every human being carries, by comparison with the reference sequence, about 10,000 missense mutations, 250 to 300 variants that lead to a gene loss, and 50 to 100 mutations that have already been associated with genetic diseases. Fortunately, in most cases, we carry these deleterious mutations in a hetero-zygous state (that is, harbouring only one copy of the deleterious allele instead of two) and, consequently, the disease does not mani-fest itself. We also know that, in different human populations, the burden of deleterious mutations is variable depending on their demographic past.

Humans are not descended from apes . . . they are apes

Another question that has obsessed humans, at least in modern times and since Darwin, is their affiliation with apes. This question has often been distorted by Darwin's opponents – and accom-panied by a long tradition of caricaturing Darwin as a monkey – for ideological reasons, in order to discredit evolutionary theories and argue that apes, as *inferior* creatures, could not compete with humans, whom *God put at the centre of His Creation*. Darwin never actually said that humans are 'descended' from apes, simply that the two species 'resemble' each other. Even though the allusion to apes in his first work is minimal, in a later book, *The Descent of Man, and Selection in Relation to Sex*, from 1871, Darwin puts forward his phylogenetic hypothesis concerning the human species, which is essentially based on comparative anatomy, underlining the great proximity between humans and the so-called 'Old World mon-keys', the Catarrhini. Nevertheless, he insists that 'we must not fall into the error of supposing that the early progenitor of the whole simian stock, including man, was identical with, or even closely resembled, any existing ape or monkey'. From that time on, as far as Darwin is concerned, humans are apes belonging to the Catarrhini, like chimpanzees, gorillas and orang-utans.

Genetics and, subsequently, genomics have not only confirmed the hypotheses of Darwin and many others about the common ancestor shared by humans and other apes, but also made it possible for us to deepen our knowledge about the human lineage. They have provided valuable data, allowing us to determine the timeline of the separation of species, the possible hybridization of some of them and the extent to which their genomes are similar, and enabling an extraordinarily precise plunge into the past and into the innermost secrets of the life of organisms and species. Just like the study of the morphological characteristics of the animals themselves – living or fossil – the study of the morphology of their chromosomes has provided information about the relationship between primate species. The karyotypes of humans and those of other great apes are very similar, but different in many respects, the most notable example being the reduction in the chromosomal number from 48 to 46 in the human lineage. Well before the era of genomics or molecular biology, comparative anatomy provided a great deal of information about the relationship between humans and other apes. In particular, it made it possible to propose an estimate of the age of separation between the ancestors of humans and those of African and Asian apes: somewhere between 15 and 30 million years ago. The techniques of molecular biology, which have developed in a spectacular way over the last few decades, have contributed new tools that prove remarkably fertile when applied to the study of the past of living things. The first molecular analyses regarding our Simian ancestry were based on immunological markers. They changed our vision of the relationship between humans and other apes. In 1963, these studies showed that humans were much closer to African apes than the latter were to Asian apes. In 1967, they dated the divergence between humans, chimpanzees and gorillas to just 5 million years ago. These results caused a great deal of surprise – and some controversy – in the scientific community.

Separation between human and non-human primates

Starting in the 1980s, researchers started using DNA to understand the relationship between humans, chimpanzees and gorillas – up until then, the order of separation between these three species had remained uncertain. To solve this question, experiments in the hybridization of DNA, a technique that compares the DNA of two species through hybridization in order to identify differences in their sequence, provided the first genetic evidence that, in two-thirds of their DNA sequence, humans and chimpanzees are much more closely linked to each other than either of them are to gorillas. Thus, in the phylogenetic tree, the divergence between humans and chimpanzees is more recent than that between their common ancestor and gorillas. Later studies at the end of the 1990s confirmed these observations by means of mitochondrial DNA or sequences of autosomes (chromosomes that are not sex chromosomes). They led to the grouping of humans with chimpanzees and bonobos, whereas gorillas diverged from the common ancestor before these three species. The orang-utan, another cousin often mentioned, is – compared with the family members mentioned here – the species most distant from humans.

The year 2005 marked the beginning of the genomic revolution in the study of the great apes, with the publication of the genome of the chimpanzee. Today, having at our disposal genomes of the different species of great apes, we can begin to explore our similarities and our differences, the timeline of the separation between the great apes and the degree of diversity within each species. On the basis of whole genomes, we know that chimpanzees and bonobos are the closest species to humans, with about 1.4% of nucleotide differences. This figure may however rise to about 5% or 6% if we take into account other types of genomic differences, such as insertions, deletions or variations in the number of copies. These figures are often used inaccurately and dogmatically, sometimes ideologically. It is important to specify they are only averages relating to the totality of the genome: in reality, the degree of divergence between species of primates varies considerably depending on which region of the genome is studied.

The latest date estimates suggest that the divergence between humans and the genus *Pan* (chimpanzee and bonobo) goes back about 4–5 million years. But these estimates have again been challenged: the speciation of *Homo* and *Pan* may have happened in several phases and the rate of mutation used to convert the percentage of divergence into time (i.e. the molecular clock) might prove weaker than was originally thought, leading to dates that are probably older (10–13 million years). As for chimpanzees and bonobos, they appear to have separated between 1 and 2 million years ago. We also know that the divergence between gorillas and the human–chimpanzee–bonobo group occurred (according to the rate of mutation and the time of generation used) between 8 and 19 million years ago, while that between orang-utans and the rest of the great apes took place 15–21 million years ago.

An ape unlike any other

So humans are apes, and they do indeed share many similarities with them. But in many respects humans are very different from their ape cousins. First of all, the genetic diversity within the human species is the weakest of all the species of apes studied, closely followed by the bonobos. Compared with chimpanzees and gorillas, humans present two to three times less diversity, even though there are almost 8 billion humans dispersed all over the world, while the area of distribution of chimpanzees and gorillas (with current numbers respectively of only 200,000 and 100,000 individuals) is restricted to Central and West Africa.

How then to explain that humans have such low genetic diversity despite their greater numbers and their wider geographical spread? The answer can be found in the first of our species: they appeared in Africa 'only' 200,000–300,000 years ago. That may seem a long time, but it is not very much on the scale of evolution. In any case, it is an insufficient length of time to allow for the accumulation of the same levels of diversity as in the other great apes. In addition, our species was probably limited to a very small number of individuals during significant periods of its evolution.

The differences we observe between the genomes of humans

and those of the other great apes, whether the genome difference is 2% or 6% in the case of chimpanzees, do not truly reflect the degree of phenotypic differentiation of the two species. It is not a problem of figures – a 'notarial' problem, as Alain Prochiantz, a neurobiologist at the Collège de France, would say. Humans share 80% of their genome with mice – and yet we are clearly quite different from them. The important thing is to see where the differences lie. If the differences from chimpanzees, however small they are in volume, are located in genomic regions involved in the development of the cerebral cortex, for example, these differences could explain the very much larger size of our brain, which is three to four times larger than that of the chimpanzee, and they may therefore be involved in the remarkable cognitive functions that our species has acquired. Small differences, large effects . . .

But these divergent nucleotides are not the only ones that may explain the phenotypic differences between humans and other great apes. Differences in structural variation, the emergence of new genes in humans and the loss of others, a great many of which are expressed in the cerebral cortex and the testicles, the variability of gene expression and epigenetic marks like the methylation of DNA – these may equally well explain the unique peculiarities of the human species when it comes to morphology, cognitive functions like articulated language, social relations, physiology and relationships with pathogens, among other phenotypic traits.

Human populations: similarities and differences

From the point of view of an individual, the differences between the genomes of two humans chosen at random, whatever their ethnic or geographical origins, are of about 3 million nucleotides, in other words, 0.1% of the genome. What about from the point of view of populations? To what extent do human populations differ among themselves? Are Africans very different from Europeans, or Europeans from Asians? Can we classify different human groups according to their genetic characteristics? These questions about the differences and the 'classifications' of human populations are probably as old as humanity itself, and, unfortunately, they have often been associated with racism.

The idea of a 'biological' concept of race goes back to a French doctor, François Bernier, who in 1684 talked about four races of humans, based especially on skin colour. In the eighteenth century, the Swedish naturalist Carl von Linné (Linnaeus; 1707–1778) classified *Homo sapiens* into categories, calling them *americanus, europaeus, asiaticus, afar* (African) or . . . *monstrosus*, in reference to the 'dwarfs' or the giants of Patagonia. During the nineteenth century, Francis Galton introduced hierarchical 'marks' (A, B, C, D, etc.): at the top of the list, he placed Europeans (in particular, the Greeks), and, at the bottom, Africans followed by Aboriginal Australians. As we shall see, the genetic data reject any scientific foundation for the concept of 'race' and show that there is no such thing as 'distinct, completely separate' human groups. These data show clearly that racism is entirely a social and ideological construct, with no biological basis.

The first studies of the biological variability of human populations began with the discovery by Karl Landsteiner in 1900 of the first genetic polymorphisms with the blood groups ABO. The red blood cells, on their surface, are carriers of antigens that may react with specific antibodies transported in the blood. Two antigens, A and B, form what is called the ABO system and define four classes of individuals: those who carry only the antigen A, those who carry only the B antigen, those who carry both (AB), and those who carry neither (O). The first population studies showed that there were indeed differences in the frequencies of these blood groups between the different human populations.

Later, other blood groups (such as MNS, Lewis and Rh) would be discovered and widely used in population studies throughout the twentieth century. The A antigen is mainly present in northern and central Europe, but also, to a lesser degree, in Africans and in Aboriginal Australians, as well as in some Native American populations in North America. As for the B antigen, the highest frequency is found in Asia, followed by Central Africa. Finally, the O antigen is mostly observed in North America, where it cohabits with the A antigen, as well as in South America, where it is prevalent in almost 100% of the Native Americans.

These methods, based on the analysis of blood, like other, similar immunological methods, were subsequently used to detect specific variants of the immunoglobulins as well as the HLA system (the human leukocyte antigen system, that is, the major histocompatibility complex or set of molecules on the surface of the cells that allows for the identification of the non-self by the immune system), which is extremely polymorphic and therefore very useful in the context of studies in population genetics. More and more polymorphisms were identified with the introduction of techniques for the electrophoresis of proteins that make it possible to separate the proteins in an electrical field according to their size and their charge. The accumulation of all these data of so-called 'classic' markers and their analysis within the context of studies in population genetics and evolution reached their height in the second half of the twentieth century, with the publication in 1994 of *The History and Geography of Human Genes* by Luca Cavalli-Sforza, Paolo Menozzi and Alberto Piazza. This work, a veritable bible of human-population genetics, used all the data from the classic markers amassed until then to gain a better understanding of the distribution of the genetic variability of our species and the history of human migration, through the study of hundreds of populations around the world.

The 1980s marked the beginning of population studies based on DNA, especially mitochondrial DNA and the Y chromosome, which make it possible to infer the past of human populations through the study of their maternal and paternal lines. The 1990s saw the consolidation of studies on these two uniparentally inherited segments of DNA, as well as the use of microsatellite markers to study human migration and population differences. With the transition to the twenty-first century, DNA studies of human populations took on a wider scope: from what was still called genetics, there was now a move to genomics. International consortiums, such as the HapMap Project, 1000 Genomes, Exome Variant Server, gnomAD or the Simons Genome Diversity Project, were set up to characterize the genetic diversity of different populations around the world.

It is worth pointing out the significance of a French project conducted by the Fondation Jean Dausset-CEPH (Centre d'Étude du Polymorphisme Humain) that has had a major impact on the study of human genetic diversity: the Human Genome Diversity Project. This project was initiated by Jean Dausset, a visionary physician and immunologist who received the Nobel Prize in 1980 for the discovery, in 1958, of the major histocompatibility complex, which has helped to identify the compatibility between donors and receivers of organ grafts. It was with the support of Jean Dausset that Howard Cann, an American paediatrician and a member of the Fondation, in collaboration with Luca Cavalli-Sforza of Stanford University, set up a sample group consisting of more than 1,000 immortalized cell lines, derived from individuals representing 54 populations that differ in their geographical location, linguistic affiliation and cultural traits. Over the last twenty-five years, although underestimated at the national level, the CEPH has acquired a high-profile international reputation. Its multi-ethnic panel has made possible the characterization of the genetic diversity of our species on an unprecedented scale.

These various international studies in genomics have provided insight into the demographic history of human populations, the way in which our ancestors admixed with other human forms like the Neanderthals and the Denisovans, and the influence of natural selection on our genome. These studies have also provided us with information that is invaluable for the clinical interpretation of the genetic variants identified in patients, and for the understanding of the genetic basis of variable morphological and physiological traits in humans, such as height or susceptibility to type 2 diabetes. But what have we learnt about the differences and similarities between human populations?

These studies have indeed been very useful. We know now that the history of human populations is marked by ancient bottlenecks, but also, in more recent periods, great expansions. Thanks to these studies, we can state that the great majority of variants observed in human populations are rare. Only those which have a frequency above 5% are considered frequent. 'Low-frequency' variants are those that present frequencies between 1% and 5%, and 'rare' variants are defined by a frequency lower than 1%. Finally, there are 'ultra-rare' variants: those that present a frequency lower than 0.01%, or have only been observed in a single individual ('singletons'). Although most variants are ultra-rare at the level of populations, at the level of an individual more than 95% of identified variants are frequent. Most differences between individuals are thus mutations that appeared early in human history, when the effective size of the population was low, and which are now present in most human populations, although at variable frequencies.

We also know now that most genetic variation is observed *within* human populations, not *between* human populations. That was the conclusion reached by Richard Lewontin in 1972 using simply a handful of classic markers. This observation, which was of crucial importance in discrediting the concept of race from a biological point of view, has been broadly supported by the many subsequent studies using a large number of human populations and whole genomes. On average, whatever the type of polymorphism analysed and the number of populations studied, 85% of genetic variability is observed *within* a population, the rest being observed *between* them.

A study published in *Science* in March 2020 analysed the results of a high-resolution sequencing of genomes from the previously mentioned HGDP-CEPH collection. The authors identified almost 70 million SNPs, about 9 million small deletions and insertions, and about 40,000 copy number variants. Once again, most of the genetic variability was observed *within* populations, with little genetic differentiation *between* populations, which form rather a gradient of genetic diversity. So why do the different human populations present so little genetic differentiation despite their wide geographical spread, whereas other species, like the Eurasian grey wolf (*Canis lupus*) or the Grant's gazelle (*Nanger granti*) of East Africa, which are much more restricted in their geographical distribution than humans, show extreme rates of genetic differentiation between populations? The answer, once again, is attributable to the recent origin of our species and the fact that humans did not evolve in completely isolated groups. Besides, no 'private' mutation, specific to a single continent or a single geographical region, has been found in 100% of the individuals of a given human population while also being absent elsewhere. This supports the idea that even if a distinction of human populations based on their genetic profiles is possible, there are no fixed differences between human groups.

The paleogenomic revolution

The study of DNA originating from fossils – the latest genomic revolution – has revolutionized our understanding of humanity's origin, the admixture events that occurred between our species and ancient, now-extinct humans, and the way in which the planet was peopled. The era of paleogenomics began in 1984 – even though it was not called paleogenomics at the time – with the publication of the sequence of a segment of mitochondrial DNA derived from the quagga, a subspecies of South African zebra that died out at the end of the nineteenth century. But it was not until the middle of the 2000s, and above all in the 2010s, that the study of ancient DNA entered its golden age. This is when paleogenomics was founded as a discipline.

Just as high-throughput sequencing overturned studies of modern populations, it also made it possible to convert the main problem of ancient DNA – its degradation into small segments (from 30 to 80 nucleotides on average) – into an asset, since in any case high-throughput sequencing requires this very fragmentation in order to sequence it. Additionally, thanks to progress both in genomic technologies and in analytical tools, it is now possible to bypass most of the other difficulties associated with the study of ancient DNA. For example, the main problems we encounter in sequencing ancient DNA are the rarity of endogenous DNA – that which really belongs to the specimen we are trying to study – and the post-mortem degradation of DNA inflicted by time. Today, techniques of enriching endogenous DNA and removing sequence mismatches make it possible to mitigate these difficulties. Fossil DNA generally presents high levels of contamination by exogenous DNA: environmental DNA (from fungi, bacteria, etc.) or modern human DNA that may have been introduced during archeological excavations or even during the experiments themselves. To estimate the rate of contamination, recently developed computational tools use, for example, the characteristics of the ancient DNA, like the excess of mutations from C to T that have come from post-mortem degradations, to help distinguish sequences of 'truly' ancient DNA from contaminants.

These dazzling technological developments have made it possible to obtain results that, barely a few years earlier, would have seemed like science fiction. For example, in 2008 we watched in awe the first sequencing of the genome of the mammoth. This feat, unimaginable a decade earlier, was made possible thanks to the extraction of ancient DNA from animal fur that had remained frozen in the Siberian permafrost for about 40,000 years. The year 2010 marked the beginning of human paleogenomics with the publication by Eske Willerslev's team in Denmark of the sequence of the first ancient human genome, derived from an individual of paleo-Eskimo origin believed to have lived in Greenland 4,000 years ago. Quite apart from the technical feat of sequencing a genome derived from a modern human at that date, uncontaminated by DNA from the experimenters, who were of course also modern humans, the interest of this study was that it brought to light a migration from Siberia to the New World about 5,500 years ago, independent of the migration that gave birth to today's Native Americans and Inuits.

Since this pioneering work, hundreds of ancient human genomes have been sequenced, some only a few hundred years old, others 45,000 years old. They derive, for example, from individuals who lived in the Upper Paleolithic, from prehistoric inhabitants of the Americas, from hunter-gatherers of the Mesolithic and Neolithic, from the first farmers of the Neolithic, up to individuals who lived in Eurasia during the Iron Age or in more recent periods. One of the oldest modern humans to have been sequenced outside of Africa is the Ust'-Ishim Man from Siberia, whose dating stretches back 45,000 years. The collaboration between the teams of Svante Pääbo at the Max Planck Institute in Leipzig and David Reich at Harvard University showed that the genome of this individual, even though he bears a greater resemblance to Eurasians than to Africans, does not share greater genetic affinities with either present-day Europeans or Asians. The conclusion to be drawn is that the population to which the Ust'-Ishim Man belonged contributed relatively little to the current genetic diversity of populations living in Eurasia.

Another example of the sequencing of ancient DNA, more notable for its popular impact than its scientific interest, is that of Ötzi, the 'Iceman'. Ötzi was discovered in 1991 by a couple of hikers in the Hauslabjoch glacier in the Ötztal Alps, near the border between Austria and Italy. He was in a mummified state: he is estimated to have died about 5,300 years ago. The most recent genetic analyses of his genome as well as of his microbiome (the genetic profile of the microbes present in the microflora), carried out in 2016, revealed a possible affiliation of Ötzi with the modern inhabitants of Sardinia. Given his genetic make-up, it is likely that he had brown eyes, belonged to blood group O and was lactose-intolerant. Moreover, the presence in his microbiome of the *Borrelia burgdorferi* bacterium suggests that he probably suffered from Lyme disease.

The Neanderthal within us

The data derived from ancient DNA have massively increased our understanding of the history of human movement and the way in which populations have adapted to their changing environments. In some cases, these studies have simply confirmed what was already known from the studies of modern DNA, that for example our species originated in Africa. In other cases, they have opened up new avenues of investigation, even running contrary to hypotheses put forward based on modern DNA. But the main surprises, the discoveries that have had the most impact in the scientific community, have come from studying the DNA of now-extinct human forms, like the Neanderthals or the Denisovans.

The relations between our species and the Neanderthals – an extinct species of the genus *Homo* that lived in Europe, the Middle East and Central Asia in a period some 40,000–230,000 years ago – have been the object of many studies and much controversy since Neanderthal remains were first discovered in 1856 in the Neander Valley near Düsseldorf in Germany. The most widely accepted idea used to be that their disappearance, which coincided with the arrival of modern humans in Europe, was the result of the physical, cognitive and technological 'superiority' of *Homo sapiens*. Some scientists were even happy to conclude that *Homo sapiens* and Neanderthals were quite different, that they did not share much and, above all, that they did not admix. Any conclusion supporting major differences between these two groups of humans was, for ideological reasons, extremely welcome.

The first genetic study of the Neanderthals, carried out in 1987 by a team led by Svante Pääbo, who was still working in Munich at the time, sequenced a small segment of their mitochondrial DNA and confirmed this hypothesis, concluding that the Neanderthals died out without contributing anything to the genetic material of modern humans. Similarly, a study conducted in 2004 by Laurent Excoffier's team at the University of Bern, a study based on the variability of mitochondrial DNA and on realistic simulations of the demographic history of the first Europeans, asserted that modern humans and Neanderthals did not admix and that, if there was any admixture, it was 'at worst' less than 0.01%. So not to worry: no genetic exchanges between the two groups!

It was not until 2010 that the first whole genome derived from a Neanderthal specimen was sequenced. The results caused a major surprise: this genome resembled more that of modern Eurasians than that of modern Africans. That was the first evidence that about 50,000–60,000 years ago, a genetic exchange occurred between Neanderthals and the ancestors of modern humans living outside of Africa today. Since then, other Neanderthal specimens have been sequenced, including three at high coverage, from the Altai mountains in Siberia and the Vindija Cave in Croatia. These genomes clearly confirmed that present-day Eurasians carry in their genomes, on average, about 2% to 3% of genetic material derived from the Neanderthals. The comparison of the genome of the Neanderthals with that of modern humans has also made it possible to estimate the date of separation of the two groups as about 550,000 to 765,000 years ago.

Humans and extinct hominins

The answer is unequivocal: our ancestors, or at least those of individuals of Eurasian origin, admixed with the Neanderthals, a phenomenon known as *archaic introgression*, and we carry the evidence in our genomes. But genetic exchanges between Neanderthals and modern humans are not limited to the transfer of genetic material from the former to the latter: a study published by Svante Pääbo's team in 2020 established that the Neanderthal Y chromosome is more likely a Y chromosome of *modern human* origin, introduced into the Neanderthal genome between 100,000 and 370,000 years ago. All these observations indicate that admixture between modern humans and Neanderthals was a common phenomenon during the evolution of the genus *Homo* (of which, today, only Sapiens remains).

But the relationship between our Sapiens ancestors and human forms that are now extinct does not stop with the Neanderthals. In 2010, the analysis of the DNA from a fossilized finger bone found in the Denisova Cave in the Altai Mountains in Siberia revealed the existence of another, previously unknown, human form. So much rested on this fossil vestige, but it was much too little: no paleo-anthropologist, however talented, would have been able to identify the species to which this remnant belonged. This is where we may appreciate the power of genetics: by sequencing the DNA, it was possible to identify a hitherto unknown archaic human, the so-called Denisovan, who probably lived in Asia about 50,000–200,000 years ago. The analysis of this genome showed that the Denisovans separated from the Neanderthals about 400,000–440,000 years ago. The Denisovans have also left traces in our genomes, which once again testifies to our ancestors' admixture with populations of archaic humans. Following a series of admixture events 25,000–45,000 years ago, Eurasians carry in their genomes less than 1% of Denisovan material, while the populations of the Pacific, notably the Papuans, carry up to 3.5%.

All these estimates of archaic introgression levels, whether from Neanderthals or Denisovans, are averages calculated across the whole modern human genome. It should be noted that some regions of our genome may have as much as 64% of archaic inheritance! Nevertheless, there was a strong selection against archaic introgression, particularly in genes coding for proteins, probably because of its deleterious effects on modern humans. The identification of regions of the human genome largely devoid of archaic inheritance (known as 'archaic deserts') makes it possible to delineate the genes and functions that may have contributed to certain features of modern humans.

For example, a Neanderthal desert has been identified around the Forkhead box P2 (FOXP2) protein, mutations in which have been associated with speech disorders. Similarly, the *AMY1* gene, which encodes salivary amylase, the enzyme responsible for the digestion of starch, is also located in a Neanderthal desert. Unlike modern humans, who carry multiple copies of *AMY1*, Neanderthals and Denisovans only had a single copy of this gene, which suggests that the production of salivary amylase for the digestion of starch conferred a great advantage on modern humans. Apart from the widespread signature of negative selection to the detriment of archaic alleles, modern humans also acquired advantageous alleles through admixture with ancient humans, a phenomenon known as *adaptive introgression*, which will be examined in depth in another chapter.

So there remains a portion of archaic humanity in us. But it is not limited to the Neanderthals and the Denisovans. There are clues suggesting other episodes of ancient admixture, including the gene flow of a 'super-archaic' group, which has not yet been identified, into the Denisovans; of the eastern Neanderthals into the Denisovans; and of archaic African humans, also still to be discovered, into modern African populations. In 2018, a study by Svante Pääbo's team of a bone fragment from the Denisova Cave in Siberia even showed that admixture between Neanderthals and Denisovans was possible. The sequencing of this specimen's DNA revealed that it came from a young girl who died at the age of about thirteen, and had a Neanderthal mother and a Denisovan father. And so we have every reason to suppose that admixture between groups of ancient humans during the Pleistocene was common whenever they encountered one another.

2.

Journeys and Encounters

What are we? So far, we have glanced at a few stages in our evolutionary genesis. But this is just the beginning of the story. Soon we will focus on what we know about the origin of the genus Homo, and above all the epic odyssey of our species, Homo sapiens, from its cradle – Africa – to the different stages of the peopling of the world and its continents. Thanks to archeology and paleoanthropology, but also, more recently, the crucial contribution of genomics, we are increasingly able to reconstruct this history and retrace the first steps of our African ancestors, following them during the remarkable trek that made them into Europeans, Asians, Oceanians and Americans as they peopled the planet. The first answer, is, then, simple: we are explorers.

In the course of these migrations, there were also encounters. Our genes bear the traces. Which means that there is another answer to the question 'What are we?' We are admixed, nothing but admixed, all of us. The verdict of genomics is definitive: we are the product of more than 200,000 years of history as a species, with around 100,000 years of journeys and many encounters in that time. What follows is our current knowledge of the subject, as it has been updated in the light of the data derived from genomics in the last couple of decades.

Over the last few decades, genetics has played an increasingly central role in our search for the origins of *Homo sapiens*. But before focusing on the *species*, we need to examine more generally the question of the *genus* and the origins of that known as *Homo*: on this point, genetics has not (so far!) provided any answers. We have seen that the human lineage, which gave birth to our species, separated from the other great apes some 7 million years ago. To understand where the different species of our lineage, that is, the *hominina* or hominins, lived during the period between the separation from the other great apes and the emergence of *Homo sapiens*, we must turn to archeology and paleoanthropology.

The human lineage and the 'great apes of small size'

Fossil remains are very clear as to the geographical origin of our ancestors. Every species of hominin living 2–7 million years ago has been found on the African continent. The first fossil that may have belonged to the human lineage dates from about 7 million years ago and was found in Africa. This is Toumaï, 'hope of life' in the Goran language, who was discovered in 2001 in Chad by Michel Brunet's team from the University of Poitiers, leading to the identification of a new species, *Sahelanthropus tchadensis*, no trace of which has been found outside Africa. Although Toumaï has a brain the size of a chimpanzee, he has a certain number of traits that link him to hominins: a flat face, the attachment of the spinal cord to the base of the skull rather than behind it, and the intermediate thickness of the tooth enamel. After Toumaï, we find, again exclusively in Africa, a series of fossil remains including *Orrorin tugenensis* (original man) in Kenya, about 5.9 million years old, *Ardipithecus kadabba* in Ethiopia, about 5.3–5.8 million years old, and *Ardipithecus ramidus* in Kenya and Ethiopia, dated to 4.4 million years ago. The affiliation of these fossil remains to our lineage remains controversial, since their physical characteristics (dental morphology, position of attachment of the spinal cord, structure of the feet) do not clearly link them to hominins.

Most fossils of hominins later than 4.2 million years ago but prior to the appearance of the genus *Homo* are attributed to the genus *Australopithecus*. The oldest known species of *Australopithecus* is *Australopithecus anamensis*, discovered in Lake Turkana and other sites from Kenya to Ethiopia, and dated to between 3.9 and 4.2 million years ago. This species is unanimously considered the oldest hominin. It is, however, less well known than another species of *Australopithecus*, namely *Australopithecus afarensis*, whose principal fossils, including those of the famous Lucy – 3.2 million years old and measuring 1.10 metres – were discovered in East Africa, and date from about 2.9–3.9 million years ago. Other species of *Australopithecus* have also been discovered, including *Australopithecus africanus* (2–2.9 million years), *Australopithecus garhi* (2.5 million years) and *Australopithecus sediba* (1.9 million years). All of them lived in sub-Saharan Africa. Although the phylogenetic relationships between them remain vague, all these species of *Australopithecus* have features in common: they are characterized by a not very large brain, taking them rather close to the other great apes, but with a set of teeth close to that of the genus *Homo* and a mostly bipedal locomotion. Another genus of hominin, *Paranthropus*, also lived in Africa 1–2.7 million years ago. It consists of three species: *Paranthropus aethiopicus*, *Paranthropus boisei* and *Paranthropus robustus*. Even though the *Paranthropus* were also contemporary with the first archaic *Homo*, it is generally agreed that they form a distinct lineage, independent of the genus *Homo*, which itself more likely descended from one of the species of *Australopithecus*.

Homo: *a 2 million-year-old story*

Homo habilis, who lived in East Africa between 2.3 and 1.5 million years ago, is the first known species of the genus *Homo*. Described in 1964, this species is characterized by a more developed brain and an ability to make stone tools – hence the name, which means 'handy man' – traits that allow it to be included in the genus *Homo*. However, *Homo habilis* presents certain morphological characteristics, like the height, the shape of the body and small teeth, that have raised doubts as to whether they truly belonged to the genus *Homo* and have led some anthropologists to classify them in the genus *Australopithecus*. They may in fact have coexisted with several species of *Australopithecus* and *Paranthropus*. Later, still in Africa, we find another species belonging to the genus *Homo*, namely *Homo ergaster* ('working man'), who lived about 1–1.9 million years ago and is sometimes described, because of similarities with *Homo erectus*, as the African *Homo erectus*. *Homo ergaster* used more elaborate cut-stone tools than their predecessors, including two representative tools: the biface and the hand axe.

It is with *Homo erectus*, 'upright man', that Africa ceases to have the monopoly of the human lineage. This is the first species of the genus *Homo* that is also found outside Africa, occupying in particular a vast territory of South-East Asia and the Far East, ranging from China to the island of Java, over a period stretching from about 100,000 to 1.8 million years ago. The best-known specimen of *Homo erectus* is Peking Man, dated to 700,000–750,000 years. His main physical characteristics are a brain volume close to that of Sapiens, a flat skull, a relatively small nose and chin, and an occipital bun. A study published in 2020 revealed that the last *Homo erectus* died out only 110,000 years ago on the island of Java – which suggests they might have encountered the Denisovans.

Before the emergence of the first *Homo sapiens*, we find other species of *Homo* spread over several periods and living in different regions of the world. It is worth mentioning in particular *Homo heidelbergensis*, who lived in Africa and Europe about 300,000–600,000 years ago; *Homo antecessor*, discovered in the north of Spain, who lived between 800,000 and 1.25 million years ago; *Homo floresiensis* (Flores Man, nicknamed 'Hobbit' because of their short height: barely a metre), found on the Indonesian island of Flores, who lived 60,000–100,000 years ago; the already mentioned Neanderthals (*Homo neanderthalensis*) and Denisovans (*Homo denisovensis*), occupying Eurasia until at least 40,000 years ago; and *Homo longi* (also called 'Dragon Man', found near Harbin in China), recently dated to 148,000 years ago and probably belonging to the Denisovan lineage, according to the paleoanthropologist Jean-Jacques Hublin, a professor at the Collège de France.

In search of the lost cradle: the arrival of Sapiens

The origin of *Homo sapiens*, the only present-day survivor of the genus *Homo*, has probably been the most controversial question in this area of research over the past forty years. They are often called 'anatomically modern humans', or simply 'modern humans', because of the anatomical characteristics that distinguish them from previous species of *Homo*: constant bipedalism, expansion of the cranium and the volume of the brain, reduction in the size of the teeth, high occipital vault, developed forehead and flat face, among others. Nevertheless, since there is no clear prototype of *Homo sapiens*, there is a sometimes-hazy distinction between anatomically modern humans and humans known as archaic. The only certainty found in the fossil remains relates to the geographical origin of *Homo sapiens*: they appeared in Africa about 300,000 years ago.

From the origins to the first exodus from Africa

Up until a few years ago, the first unquestioned representative of *Homo sapiens* was Omo Kibish, found in Ethiopia, who lived about

195,000 years ago. But older specimens presenting physical charac-teristics very close to those of *Homo sapiens* have recently been discovered: for example, the specimen discovered at Jebel Irhoud in Morocco by Jean-Jacques Hublin and his team, which has been dated to about 300,000 years ago. It had previously been accepted that *Homo sapiens* first appeared about 200,000 years ago: now its origin has been pushed back by 100,000 years!

Studies conducted over the last few years regarding the exodus of *Homo sapiens* from Africa have sprung a few more surprises on us. It was long believed that modern humans left Africa for the first time some 90,000–120,000 years ago, first reaching the Middle East, as demonstrated by fossil remains dating from that period found in the Qafzeh and Skhul Caves in Israel. But in 2018, researchers published a study based on an upper jawbone of *Homo sapiens* discovered in Misliya Cave, also in Israel. These fossil remains correspond to the oldest member of the family of *Homo sapiens* identified outside Africa. After analysis, dating estimated its age as being between 177,000 and 194,000 years, pushing back by about 70,000 years the first exodus of *Homo sapiens* from Africa. Until then, the fossil remains of *Homo sapiens* found in other parts of the world had all been dated as less than 60,000 years.

What this set of data suggests may be summarized as follows: the migrations associated with this first exodus were not followed by the permanent settlement of the first humans outside Africa. The oldest fossils that have recently been discovered do indeed reflect an expansion of the African geographical area of Sapiens, given the position of the Middle East as a geographical corridor. But the prin-cipal phase of human dispersal took place much later. Genetics has confirmed that a second wave of migrations began 60,000–80,000 years ago, leading Sapiens out of Africa and reaching, on the one hand, the Middle East, Europe and Eastern Asia, and, on the other hand, southern Asia by a coastal route. But questions of the number, the geographical origin and the routes of migration taken by the first humans leaving Africa are far from being resolved, and there is no shortage of controversies.

A single ancestral origin, or several?

The theory that *Homo erectus* set out from Africa has long been accepted. But the way in which *Homo sapiens* spread throughout the planet has been widely debated in the scientific community, and especially so at the end of the last century. Based always on fossil remains, paleoanthropologists proposed different models to explain how *Homo sapiens* evolved from its predecessor, *Homo erectus*, and above all to determine the geographical region where this transition (or these transitions) took place. On this matter, there was much heated debate between, on the one side, the champions of monogenism, who believed that all human groups shared a common ancestor, and, on the other side, the adherents of the polygenist models, who considered that the different human populations occupying the planet today – the human 'races', as was said at the time – had quite distinct origins and in no way shared a common ancestor, in any case not a recent ancestor, let alone one whose origins lay in Africa!

Among the polygenists, the influential Carleton Coon, also known for his racist views, proposed at the beginning of the 1960s the candelabra model, according to which today's African, European, Asian and Oceanian populations are the result of the independent and separate evolution of local *Homo erectus* towards *Homo sapiens* on each of the continents. Unsurprisingly, Coon championed the idea that Europeans were the first to evolve from Erectus to Sapiens, the last being the Africans, classified as the population closest to the 'primitive' state of the genus *Homo*. The candelabra model was widely accepted by the anthropological community for many years, even though it involved the somewhat complicated idea of a parallel and convergent evolution of the traits characterizing anatomically modern humans.

To mitigate these shortcomings and get around Coon's racist biases, other anthropologists proposed an intermediate version of this model, the *multiregional model*: without challenging the idea of a parallel evolution of Erectus towards Sapiens everywhere in the world, this model hypothesizes that there may have been genetic exchanges between the different continental populations, which would explain why they present common phenotypic traits. The multiregional model is mainly associated with Milford Wolpoff, who proposed it in 1984, although Franz Weidenreich had already hinted at it in 1947. This model also accepts a regional continuity on each continent, in other words that present-day Europeans share links with ancient Europeans, such as the Neanderthals, and similarly for the other populations of the globe.

Unlike the polygenist models, monogenist theories maintain that the transition from Erectus to Sapiens took place on a single continent. It is within this context that in 1988 anthropologists Chris Stringer and Peter Andrews proposed the *Out of Africa model*, according to which all current populations of *Homo sapiens* derived from a single group of Sapiens that appeared first in Africa and later left its homeland and totally replaced the other populations of archaic humans living on all the other continents.

In 1989, Gunter Brauer proposed a variant of this model, the *assimilation model*, which suggested that while dispersing across the world from Africa, *Homo sapiens* did not entirely replace the other species of humans living outside Africa, but on the contrary hybridized with them.

The limitations of archeology and paleoanthropology

Even though all these models concerning our origins are based on fossil remains found in different places around the world, or perhaps for that very reason, it is difficult to decide between them simply on the basis of archeology or paleoanthropology, since they are all different interpretations of the same facts. While archeological research has certainly managed to successfully elucidate the diffusion of cultures, its ability to determine whether the diffusion of a culture happened because of the movement of people or the spread of ideas is very limited. As for paleoanthropology, which seeks to classify fossil remains based on bone morphology, it may inform us as to the movements of populations, but it is also limited by the preservation of fossil remains and by biases caused by the fact that not all geographical areas have been excavated, which means that the data are incomplete and cannot provide a complete picture.

On the other hand, genetic analyses of both current and ancient populations make it possible to directly establish genealogical links between humans, by providing a high-resolution image of migration routes and the epochs of admixture between different groups. This incredibly powerful tool came along at just the right time to answer a number of questions that other disciplines could not address.

The first answers arrived towards the end of the 1980s. Starting from each of the above-mentioned models, it was possible to make estimates of the quantity of African ancestry detectable in today's populations. At the two extremes, we find, on the one side, the candelabra model, which suggests that populations of Sapiens outside Africa will have no African ancestry, while, on the other side, the Out of Africa model estimates 100% African ancestry in every population in the world. As for the intermediate models, they estimate that 'most' of the gene pool of non-African populations will be either of local origin in each continent, as maintained by the multiregional model, or of 'mostly' African origin, as supposed by the assimilation model, with some traces of other species, Neanderthals in Europe, for example. But these models can be tested perfectly well by genetics – and so we are now able to finally decide between them.

Homo sapiens *is African*

The first genetic data on the origin of our species came from the analysis of a small molecule, mitochondrial DNA (mtDNA), by Allan Wilson's team at the University of California in Berkeley in 1987. By analysing the variability of mtDNA in different populations in the world, Wilson and his colleagues showed that African populations present more variability than other populations spread around the globe – almost double, in fact. Above all, they established that the roots of these latter in the phylogenetic tree are in Africa. These results provided the first genetic evidence supporting the Out of Africa model.

During the following decade, a large number of populations were studied thanks to increasingly advanced analyses of mtDNA. The evidence piled up, confirming the first results: all human populations share a common origin in Africa. In addition, this origin is quite recent, if we trust the hypothesis of the molecular clock. This hypothesis is based on the idea that mutations accumulate at a constant speed, and as such we are able to use the 'amount' of genetic variability as a measure of time. Thus, on the basis of the variability of mtDNA, the appearance of our species was dated to between 150,000 and 200,000 years ago.

However, mtDNA reflects only part of our history: that of our maternal ancestors. It is thus quite possible that the phylogenetic tree inferred from a single marker, like mtDNA, is not representative of the overall genomic pattern of human history. To submit the different models of the history of our species to a more rigorous test, we must call on other parts of our genome. That is why, in the 1990s, there was a huge increase in the number of studies based on the analysis of the Y chromosome, inherited through the paternal line. The conclusions confirmed once again that our species originated uniquely, and recently, in Africa. Nevertheless, these conclusions relied again on only two genetic markers: the information obtained was infinitesimal compared with that contained in our entire genomes . . . which have 3 billion bases!

Since the start of the twenty-first century, thanks to technological progress in genomics, whole-genome studies have become widespread. In the 2000s, it was the analyses of microarrays that took precedence. Microarrays make it possible to examine in a single experiment of hybridization the presence of millions of genetic variants, the *single nucleotide polymorphisms* or SNPs mentioned in the previous chapter. Data from studies using microarrays, like the HapMap Project, have clearly confirmed that our species originated in Africa. They have also shed light on the genetic links between human populations, the genetic structure observable within continents, and the levels of admixture between different groups.

Microarrays were able to measure the genetic variability of human populations at a much higher level of resolution than mtDNA or the Y chromosome. But they also have their limitations. To the extent that they examine exclusively the SNPs discovered in a restricted number of individuals from a few populations, they cannot reach an overall, unbiased view of the genetic diversity of world populations, which would be necessary if we wanted to infer a faithful picture of their demographic past. These limitations have been overcome with the arrival of high-throughput sequencing at increasingly lower costs. With these techniques, we have been able to tackle, without bias, a large number of questions about the origins of our species, the demographic history of human populations in terms of their divergence, the change in size of these populations over time (due for example to bottlenecks, founder effects, etc.), and levels of genetic exchanges both between human populations and between humans and now-extinct hominins.

What do we know today about the origin of humans and how they dispersed throughout the world? We can now turn to genomic studies integrating data derived, on the one hand, from modern populations – the 1000 Genomes Project is a representative example – and, on the other hand, from archaic hominins such as the Neanderthals and the Denisovans. What emerges unequivocally from the currently available data is that modern humans appeared in sub-Saharan Africa at least 200,000 years ago. One important detail needs to be noted: among human populations, African populations present the highest level of genetic diversity in the world. Moreover, the diversity of non-African populations decreases the further we get from Africa. These facts attest to the existence of founder effects in the course of the migrations of these populations across the globe, according to the so-called 'serial founder effects'. And so the genomic data of modern populations also clearly support the idea that *Homo sapiens* is African.

Out of Africa

Both archeological and genetic data show that Sapiens was born in Africa and remained there for at least 100,000 years before deciding to leave. The consensus, from the genetic point of view, is that they started to disperse across the world 60,000–80,000 years ago, quickly reaching South-East Asia, Australia, Europe and Eastern Asia. Humans later reached more remote destinations like the Americas about 15,000–35,000 years ago, then the remote islands of Oceania, where they settled more recently, only 1,000–4,000 years ago. The following chapters will discuss in detail the way in which the peopling of the different continents unfolded.

One important question remains unresolved: when and how did the first Sapiens leave the African continent? In 1994, the anthropologists Marta Lahr and Robert Foley advanced the hypothesis that there were multiple dispersals of modern humans out of Africa during the Pleistocene. Based on fossil remains and archeological findings, they suggested that the first *successful* dispersal out of Africa was accomplished via a southern coastal route, leaving East Africa, following the coasts of southern Asia, and finally reaching Australia about 50,000 years ago. Then along came genetics. The first data derived from mitochondrial DNA and the Y chromosome in the 2000s pointed rather towards a single coastal exodus of modern humans about 60,000 years ago. This hypothesis was strengthened with the arrival of studies using microarrays. Nevertheless, other explanations remain in the running and the subject is still wrapped in controversy. Recent studies using whole-genome sequencing have shown that all contemporary non-Africans are descended from a single population that left Africa at least 60,000 years ago. True, the modern inhabitants of Papua New Guinea present genomic traces of an older dispersal from Africa, about 120,000 years ago, but this still remains to be confirmed and all the evidence suggests that it made only a small contribution to the diversity of most present-day Eurasian populations.

Ancient DNA has also provided valuable information as to the exodus of our species from Africa. As we saw in the previous chapter, the genomes of all individuals of non-African origin contain about 2% of genetic material of Neanderthal ancestry, which suggests that admixture with Neanderthals took place not long after the exodus from Africa, probably in the Middle East. This is compatible with the model of a single exodus of humans from Africa. More generally, the data from ancient DNA have made it possible to refine the different models proposed to explain the origins of our species. The model most supported by genomics today is the assimilation model: our species, born in Africa, set off to populate the rest of the world and 'almost' replaced the totality of the local populations of archaic humans that it encountered.

The long and rich African history

It would obviously be misleading and grotesquely reductive to see Africa as nothing but the birthplace of our species, and ignore the subsequent African history of *Homo sapiens*. In fact, the populations residing on that continent had an extremely complex demographic history even before their dispersal to the rest of the world. Not only does Africa present the highest level of human genetic diversity on the planet, it is also characterized by a great diversity of languages, cultures and phenotypes. More than 2,000 different languages are spoken in Africa, almost a third of all the languages in the world. Moreover, a wide range of modes of subsistence are practised by African populations, including various forms of agriculture, pastoralism, hunting and gathering. In addition, these populations live in remarkably varied environments: Africa contains the largest desert in the world and the second-largest tropical rainforest, as well as savannahs, marshlands, high mountainous regions and much more. Sapiens was initially shaped in this world.

Different cradles for Homo Sapiens?

But, before examining the history of African populations, we must return to the question of the origins of *Homo sapiens*. Although it is beyond question that the origin of human beings is located in Africa, much still remains to be discovered: for example, the precise place or places in that vast continent where we first appeared. As with the origin of humans in general, different models have been developed to retrace the genesis of our species *within* Africa and to determine its place of birth: the model of *African multiregionalism*, the model of *a single origin with local extinctions*, the model of *a single origin with regional persistence* and the model of *archaic hominin admixture in Africa*. Starting with each of these models, we can also formulate estimates of the morphological, genetic and archeological findings that they should lead to, these estimates being different and specific to the model. But we do not always have a clear and reliable answer that everyone agrees on. It should be said, though, that the integration of data deriving from different disciplines does rather suggest, in accordance with the *multiregional model*, that *Homo sapiens* does not have a single geographical origin in Africa, but may have evolved, in various places on the continent, from older forms that subsequently admixed between themselves, and with archaic humans.

Ancient admixture

In the last few years, genetic data have provided us with more and more factual evidence in favour of the idea that the current genomes of African populations bear traces of ancient admixture with archaic humans, as we saw with the Eurasians. Contrary to what was initially thought, traces of Neanderthal introgression have also been observed in some populations in sub-Saharan Africa. They bear witness to the recent admixture of these populations with Eurasian populations, which themselves have Neanderthal ancestry.

More surprisingly still, different studies have detected in present-day African populations genetic material deriving from another hominin. Since this material is not of Neanderthal, let alone Denisovan origin, it may derive from now-extinct archaic humans unknown to paleoanthropology. Archaic introgression has been estimated to be as high as 2% to 8%, depending on the population studied and the method used, although the degree of archaic admixture in Africa with this unknown hominin remains controversial. These observations nevertheless underline the value of the genomic approach. Even in the absence of archeological or paleoanthropological evidence, especially in Africa, where environmental conditions are not ideal for the preservation of fossil remains, this approach has allowed us to 'resuscitate' ancient human forms that are now extinct. A large number of ongoing studies are attempting to resolve this thorny question.

The genomic history of hunter-gatherers

Genomic studies, relying both on the DNA of modern populations and on ancient DNA, have also provided insight into the history of the peoples of Africa. One of the most surprising observations concerns the very remote period of the separation of certain groups of hunter-gatherers. Agricultural practices spread through Africa with what are known as the Bantu expansions, which we will discuss later. After this expansion of agriculture, only a few populations conserved their original mode of subsistence based on hunting and gathering, to different extents. Of particular note are the rainforest hunter-gatherers of Central Africa and the Khoe-San populations of southern Africa, as well as the Hadza and Sandawe of East Africa. The most ancient population split we know of, among all human groups living on earth, is the one separating the Khoe-San from other populations. It has recently been estimated, taking into account new mutation rates, to have happened about 260,000–350,000 years ago! These very ancient times suggest that the separation of the ancestors of the Khoe-San peoples of southern Africa took place soon after the birth of *Homo sapiens* itself. The second oldest separation is that of the rainforest hunter-gatherers: it is estimated at about 135,000 years ago, but the dating is uncertain, and this separation might have been close to that of the Khoe-San.

Once again, it is genetics that has made it possible to reconstruct the extent of the original territory of the hunter-gatherers. Various studies have reported genetic relationships between the Khoe-San hunter-gatherers of southern Africa and the Hadza and Sandawe of East Africa. These observations have led researchers to suggest that, before the great demographic changes that came about as a result of the expansion of agriculture across sub-Saharan Africa, the territory of today's hunter-gatherers, as represented by the Khoe-San, was much larger and stretched as far as East Africa. This hypothesis is also supported by the linguistic data, since the Khoe-San and some East African populations have one notable thing in common: the use of click languages.

The spread of agriculture and Bantu languages

Whether it is the rainforest hunter-gatherers, the Khoe-San, or the Hadza and Sandawe, the current populations of these groups are represented only in pockets, bringing together the survivors of two major cultural events in the history of Africa: the spread of the Bantu languages and the emergence of agriculture. On the basis of linguistics and archeology, we know that some 4,000–5,000 years ago, Bantu-speaking peoples, until then hunter-gatherers living in a region situated between Cameroon and Nigeria, began to master agricultural practices. Their language and lifestyle, based on agriculture, sedentism and the mastery of iron, gradually spread all over sub-Saharan Africa. Today, although they are distributed over a surface area of 500,000 square kilometres, most of the populations of sub-Saharan Africa speak one of the 500 languages belonging to the Bantu family.

Archaeological evidence indicates that agricultural practices spread over large distances in an odyssey that lasted several millennia. But, until recently, one question remained to be cleared up. Simply on the basis of archeological data, it is not easy to determine what actually spreads: is it exclusively cultural *practices* linked to agriculture – which would then be *cultural* diffusion – or is it the populations themselves, who practised agriculture and who migrated – which would be a *demic* diffusion? Once again, genetic studies offer a unique opportunity to study the demographic effects of agriculture by comparing genetic variation within agricultural populations with the groups of hunter-gatherers that survived. Verdict: both the initial studies, based on the analysis of mitochondrial DNA or the Y chromosome, and more recent, whole-genome studies point to the great genetic similarity of groups speaking Bantu languages, even though they are geographically distant, compared with populations from other linguistic families. In this way, the genetic data unequivocally confirm the model of demic diffusion of both the Bantu languages and agriculture, making the Bantu expansions the greatest population movement in the history of Africa.

Even though there is no doubt about the fact that the Bantu peoples migrated, the question of which route these peoples took remained unresolved until recently. An initial theory, known as the *Early Split*, asserted that the movement divided into two from the start, when the Bantu farmers left their original cradle, between what is now Cameroon and Nigeria, one group going eastwards, the other southwards. The rival hypothesis, called the *Late Split*, suggested that these peoples first crossed the equatorial rainforest – in what is now Gabon – before dividing into two migratory flows, one going south and the other towards East Africa. Here too, genetic studies seem to have settled the matter. The most recent studies, including those led by Etienne Patin (Institut Pasteur/CNRS), reveal that the Bantu-speaking populations of eastern and southern Africa are genetically closer to the populations of the south than to those north of the equatorial rainforest. These data clearly endorse the Late Split theory, showing that the Bantu farmers first crossed the equatorial forest and then followed their migratory routes to the east and south of sub-Saharan Africa, where they encountered local peoples native to these regions.

The diffusion of animal husbandry

Population movements associated with new modes of subsistence and nutrition are not limited to the Bantu expansions or the diffusion of agriculture and the use of iron. Archaeological data also suggest that an economy based on raising livestock, especially sheep, was introduced into southern Africa from East Africa, probably reaching Zambia and Zimbabwe about 2,000 years ago. However, as with the Bantu expansions, it has not been possible to establish from archeological artefacts alone whether the spread of pastoralism was associated with a demic diffusion of the pastoral populations or a diffusion of the practice itself. Genetic studies based on both current populations and ancient remains of the Khoekhoe (herders) and San (hunter-gatherers) of southern Africa have detected a mixed East African/Eurasian genetic component in all the Khoe-San groups. It is interesting to note that this component is present at particularly high frequencies, of the order of 23% to 30%, in the Khoekhoe Nama, whose date of admixture has been estimated at about 1,300 years ago. The conclusion drawn from these studies is that animal husbandry was brought to southern Africa through the migration of a group of people from the east of the continent, who subsequently integrated with local groups of hunter-gatherers in southern Africa, giving birth to the ancestors of today's Khoekhoe herders. There were many admixture events between groups of farmers or shepherds migrating into different regions of Africa and local groups of hunter-gatherers. A later chapter will be devoted to this subject.

Admixture in East Africa

It is in East Africa that the complexity, the demographic, linguistic and cultural richness, of Africa is at its greatest. Indeed, the emergence of agriculture in the continent is not linked exclusively to the Bantu expansions, since agricultural practices in East Africa seem to have developed independently in the eastern Sahara/Sahel region about 7,000 years ago, and in the highlands of Ethiopia between 4,000 and 7,000 years ago. Today, the farmers of East Africa speak Afro-Asiatic and Nilo-Saharan languages, and, as in other regions of Africa, modes of subsistence based on hunting and gathering have been almost completely replaced. In addition, genetic data indicate that the populations of East Africa present very high levels of admixture with other populations of varied origins. Indeed, genetics shows that the region was strongly influenced by different migrations. During a migratory event known as *Back to Africa*, populations from the Levant contributed a Eurasian genetic component to the people of East Africa in the course of the last 3,000 years, while migrations of farmers from the western part of Central Africa provided the 'Bantu' component over the last 2,500 years.

Very few populations have preserved the autochthonous gene pool of East Africa from before the arrival of the Eurasian migrations or the Bantu expansions, but some that have are the populations of Nilotic shepherds in South Sudan, like the Dinka, the Nuer and the Shilluk, or the Hadza hunter-gatherers of Tanzania. The genetic data also show that there were cultural and genetic exchanges in both directions, east to west and west to east, between groups of nomadic farmers in the whole belt of the Sahel, the corridor between the equatorial rainforest to the south and the Sahara Desert to the north. However, the genetic data of East African populations remains sparse: a greater sampling of the populations of this region would provide a better understanding of the complexity of this key territory in the history of humanity.

North Africa, crossroads of migrations

Most genetic studies have focused on sub-Saharan Africa. North Africa has not been studied to anything like the same extent, even though it has a rich, complex and fascinating history. Thanks to their strategic position, the populations of North Africa are the result of a great mix of migrations. They occupy a territory situated at the crossroads of three continents. In the south it is bordered by the Sahara Desert, which acts as a barrier between it and the rest of the African continent; in the north it opens onto the Mediterranean basin, which has facilitated trade with seafaring civilizations from Europe; and finally, it is linked to the Middle East by the Arabian Peninsula and the Sinai, allowing constant migrations over land. In addition, North African populations speak two languages from the Afro-Asiatic family: Arabic, introduced into the region from the Middle East during the spread of Islam, and Berber, a language group that encompasses many languages and dialects. We should add that human occupation in North Africa dates back at least 130,000–190,000 years – not to mention the presence in Morocco of Jebel Irhoud Man, who bears characteristics very close to those of modern humans, and who lived about 300,000 years ago.

What emerges from the genetic studies are some remarkable lessons. Of particular note is the work of David Comas from Pompeu Fabra University in Barcelona, which has above all revealed high levels of admixture in current North African populations. Within them, individuals of the same geographical origin are found in different genetic groups, without any clear demographic, ethnic or geographical classification. Genetic data derived from today's populations, combined with information derived from ancient DNA, show that most North African populations have been strongly influenced by a component from the Middle East, associated with the Arab expansions, which had a major impact on the eastern part of North Africa, and a Maghreb genetic component, representing the local ancestry of the region, which is found at high frequencies in the western part of North Africa.

To a lesser extent, genetic data have revealed a recent influence of sub-Saharan Africa in North Africa, especially in the western part. This has been associated with the slave trade in the region during the period of ancient Rome. Traces of admixture with European populations have also been detected. As for the Berbers, considered the autochthonous peoples of the region, genetics shows that they also consist of very heterogeneous groups. They, too, went through a major period of admixture – of very variable intensity, depending on the groups concerned – especially following the Arabization of North Africa, which led to a strong dilution of their common, initial genetic ancestry.

A highly admixed Europe

If there is one continent that has been studied from every angle – and genetics is no exception – that continent is Europe. Archeologists, paleoanthropologists and geneticists have scrutinized the history of European populations in much greater depth than those of any other continent. According to the archeological data, the first anatomically modern humans in Europe date back at least 42,000–45,000 years. It was during this period that the first Eurasians admixed with Neanderthals. It is also quite clear that these first Eurasians left very few genetic traces in today's populations. The oldest genomic data from a modern human in Europe come from the individual known as Oase 1, discovered in what is today Romania. He is 37,600–41,600 years old and had a direct Neanderthal ancestor during the previous four to six generations.

Four-stage peopling process

Tallying the information from different disciplines concerning the movements of populations, we can divide the history of Europe into four great stages. The first corresponds to the initial peopling of the continent from the Middle East, about 45,000 years ago. The second covers the phase of the repeopling of Europe after the last Ice Age, which took place about 15,000–26,000 years ago. This Ice Age led to a steep population decline, to the benefit of several southern refuges where the climatic conditions were much more favourable, like the Franco-Cantabrian region – between what is now France and Spain – or the area around the Black Sea. At the end of the Ice Age, populations of hunter-gatherers spread across Europe from these refuges. The third stage stems from the arrival of agriculture in Europe. About 11,000 years ago, a new way of life based on livestock, agriculture and sedentism – characteristic of the Neolithic period – began to develop in several regions of the Fertile Crescent. From this region, most likely from Anatolia, a farming culture spread to Europe, reaching as far as the Iberian Peninsula, Scandinavia and the British Isles sometime in the last 6,000 to 8,000 years. Finally, the last great cultural change to leave a deep mark on the history of Europe was the migration from the east at the end of the Neolithic period and the beginning of the Bronze Age. This migration is attributed to the pastoralist nomads of the Pontic-Caspian steppe, belonging to the Yamnaya culture, who reached Europe about 4,500 years ago and spread the Indo-European languages spoken in Europe today, but this remains a subject of debate.

Despite the extent of study focused on Europe, several questions have remained for a long time unresolved, and some are still the object of in-depth research. What is the relative contribution of these different migrations to the current genetic make-up of Europeans? Were some regions more affected by these migratory movements than others? Was the arrival of agriculture in Europe the result of a cultural transmission of techniques or, rather, a genuine movement of populations? As seen in the preceding chapter, it is now possible to test these hypotheses by analysing the genetic variability of today's populations. What do these studies reveal? Based on the data obtained from the so-called classic markers, but also from the Y chromosome, the first thing that was observed is the presence of gradients of frequency in the direction south-east to north-west. Initially, this observation was interpreted as supporting the model of demic diffusion of agriculture during the Neolithic Age – this is notably the position long held by the school of Luca Cavalli-Sforza. Nevertheless, other population movements in the same direction, but occurring at other periods, could have left similar genetic traces. In addition, simulation studies showed that such gradients of frequency can be obtained even in the absence of any spatial expansion.

Tripartite admixture

Was the diffusion of agriculture in Europe demic or cultural? It is ancient DNA that finally settled the controversy. Over the last decade, especially in the last five years, studies of ancient DNA in Europe have multiplied. They have shown that today's European populations are the result of intense admixture between three genetic components, associated with different ancestries: the hunter-gatherers of western Europe in the Mesolithic, farming peoples from Anatolia in the Neolithic and migrants from the steppes of Central Asia, associated with the Yamnaya culture. The proportions of these components are very variable depending, on the geographical region studied. For example, analyses of DNA dating from different periods, from the Mesolithic to the present day, show that the 'hunter-gatherer' genetic component began to be heavily diluted by the arrival of farmers from Anatolia and the assimilation of the local hunter-gatherers, about 8,500 years ago. These results constitute a blunt demonstration that the Neolithic lifestyle was spread across Europe by the migration of people, as Cavalli-Sforza predicted, rather than by the diffusion of techniques. The Neolithic lifestyle also seems to have contributed towards an increase in the size of populations, as suggested by the estimates of the effective size of the populations that have been generated from genomic data.

Once again, it is genetics that has made it possible to establish that the Neolithic migrants from Anatolia did not completely replace the populations of hunter-gatherers. About 4,500 years ago, almost all the populations of Europe were already admixed and possessed both genetic components, with between 10% and 25% of ancestry derived from hunter-gatherers from the West. Today, the 'hunter-gatherer' component remains the least represented. It is in the populations of northern Europe that its frequency is highest. Concerning the genetic component attributable to farmers from the Middle East, the highest frequencies are found in the populations of southern Europe, in the Sardinians in particular, while the lowest are observed among the populations of northern Europe.

Finally, it is thanks to studies of ancient DNA that it has been possible to describe the massive influence on the genetic make-up of Europeans of the migrants coming from the steppes of Central Asia affiliated with the Yamnaya culture during the Late Neolithic and the Bronze Age. These populations from the East were themselves the descendants of various groups of hunter-gatherers from eastern Europe and the Caucasus region. The genomic data indicate that the component from the steppes was already present in southern Europe 6,000 years ago, in northern Europe 5,000 years ago and in central Europe about 4,500 years ago. The steppe ancestry is most frequent today in central and northern Europeans, but weak in populations of the South.

Geographical and genetic maps

Over the last 10,000 years, as we have seen, Europe has been the setting for various migratory events followed by a marked admixture of peoples. This makes today's Europeans highly admixed populations, carriers of genetic material taken from different groups of migrants – to which we must also add . . . the Neanderthal component! All these events led to the distribution of the genetic diversity that we observe in today's Europeans, and which follows a model of isolation by distance: geographical and genetic distances are correlated, and, in turn, the closer that two populations are geographically, the more similar they are from a genetic point of view. In other words, the current genetic diversity of Europe cannot be dissociated from geography and shows a decreasing gradient of diversity the closer we get to the northern latitudes.

In a 2008 study, John Novembre, who was working at the University of California in Los Angeles at the time, showed that, with genetics, the geographical origin of a European individual can be estimated with a precision of between 500 and 800 kilometres. The correlation between genetic and geographical diversity in Europe is such that it can even be detected on the scale of small geographical areas like Switzerland, Finland, Iceland or Great Britain.

Borders, admixture and isolation in Europe

This very marked correlation between geography and genetics in today's European populations also makes it possible to reconstruct the migratory past of specific regions or countries. For example, a study based on the genomes of more than 2,000 individuals from the United Kingdom, published in *Nature* in 2015, showed that the genetic contribution of the Anglo-Saxon migrations to the current populations of south-east England is less than 50%. It also hypothesized that there were population movements from continental Europe to south-east England between the end of the Mesolithic and the beginning of the Roman period and suggested that there must have been genetically differentiated subgroups in the non-Saxon regions of the United Kingdom, rather than a base population of Celtic origin.

France has long been conspicuously absent from studies of genetic diversity. However, in 2020, Christian Dina's team at the University of Nantes analysed the genetic profiles of more than 2,100 individuals originating from different regions of France. It was revealed that French genetic diversity also follows a model of isolation by distance. Different genetic components can be observed, their proportions varying greatly at the level of certain geographical barriers. For example, a clear genetic division separates the populations of northern France from those of the south, with a genetic barrier situated along the Loire River, which for a long time was a political and cultural border between the kingdoms or counties of the north and the south. Similarly, though to a lesser extent, the Garonne and the Adour contributed to the differentiation of the populations on either side of these rivers, consequently playing an important role as a barrier preventing the mixing of the populations. Overall, the regions that show the strongest genetic differentiation are Aquitaine and Brittany. This observation is consistent with the cultural and political history of these territories, which led to their populations being more isolated compared with others. It should however be emphasized that, even though historical, cultural and political borders seem to have shaped the genetic structure of modern France, its historic population remains, at least from a genetic point of view, quite homogeneous.

At this point, it must be noted that the levels of genetic differentiation between populations are always relative and depend above all on the scale of the phenomena considered. True, the populations of Aquitaine and Brittany are more 'differentiated' compared with the populations of other French regions, but the extent of these differences must be considered in a broader context: within a broader scope of analysis, they appear to be only minor nuances. One of the aims of population geneticists is precisely to search out these differences and evaluate their relevance, so as to use them to reconstruct the history of a people or geographical region. By using the right means and the required level of resolution, we could even see differences between Montpellier and Toulouse, if we wanted! But here too, it is not a matter of an accounting or 'notarial' logic. Not all differences are significant: it all depends on scale and how scientifically interesting the question is.

Following the same argument in the context of European differentiation, we note that some populations stand out from the crowd. We can observe several notable discrepancies compared with the general model of isolation by distance. The Finns, the Sardinians, the Basques and the Ashkenazi Jews are good examples of this: their demographic, geographical or cultural history has led to greater genetic differentiation . . . *relative to other populations*. The lack of genetic differentiation of some populations compared with their neighbours may also be of relevance. For example, the Hungarians, even though they speak a Finno-Ugric language, are genetically similar to their geographical neighbours who speak Indo-European languages. This observation sheds light on a case of linguistic replacement that was not followed by genetic replacement. This is a phenomenon known as 'elite dominance': a few individuals constituting an elite, often from elsewhere, impose a new language but do not admix with the local population.

The 'Asias': lands of contrasts and encounters

Although Asia contains a rich cultural and linguistic diversity, the genetic structure of the populations across the continent remains, unfortunately, one of the least studied. Moreover, the Asian continent is so vast, containing so many geographical barriers and strongly differentiated cultural entities, and such a variety of environmental conditions, that it is difficult to sum up the history of the peopling of Asia, composed as it is of several independent and complex migratory events. Early genetic studies supported the idea of a single coastal migration via southern Asia about 60,000–70,000 years ago, which lay behind the initial settlement of South-East and East Asia, followed by expansions northwards around the Tibetan plateau. However, populations of Papuan origin carry signatures, at a level of 2% of their genome, attributed to the first exodus from Africa via the Middle East about 120,000 years ago. This exodus, as we have seen, was a dead end and contributed very little to the genetic diversity of present-day populations outside Africa. As for the coastal migration, it probably followed the coasts of the Arabian Peninsula, the Persian Gulf and India, and very soon, about 58,000 years ago, reached South-East Asia and Sahul – the name given to the continental landmass from which Australia, Tasmania and New Guinea have emerged over the last 10,000 years. Studies of the populations of South-East and East Asia have also revealed a strong correlation between genetic diversity and latitude, which has been interpreted as the result of a series of founder effects as human populations moved from the south-east to northern Asia.

In addition, the ancestors of Eurasians diverged from the populations of South-East Asia and Sahul about 57,000 years ago, then diverged once again about 42,000 years ago, giving birth to the

Western Eurasians – of which the Europeans are part – and the Eastern Eurasians. Thus, the genetic data support a single 'successful' exodus from Africa towards Eurasia, in a movement that subsequently split into two great waves of migration in Asia, one going south and the other north.

The history of the populations of continental Asia, from a cultural as well as geographical, linguistic and archeological point of view, can be divided into several 'Asias', each corresponding to quite differentiated entities or genetic components. Despite the overall scarcity of studies of the populations who live on the Asian continent today, the genetic data show that we can divide Asian populations into several large genetic groups which correspond, with a few exceptions, to the linguistic families of the region: the Altaic, Sino-Tibetan/Tai-Kadai, Hmong-Mien, Austroasiatic, Indo-European and Austronesian, among others.

There are exceptions. For example, the genetic components associated with isolated populations or populations with specific demographic histories, as in the case of the Mlabri of Thailand and Laos, a group of nomadic hunter-gatherers of whom there are barely 200 individuals remaining today, or the 'Negritos' of the Philippines, Malaysia and India, populations of small size and dark skin. A recent study, based on the sequencing of the genomes of more than 1,700 individuals from 219 populations in Asia, established that the different groups classified under the term 'Negritos' are genetically closer to their geographical neighbours, whether in the Philippines, Malaysia or India, than to other groups of Negritos. This suggests that the phenotypic traits that characterize them, like dark skin and small size, are the result of adaptation to their environment and way of life rather than indicative of a common origin.

What genes reveal about Asian history

As with the history of Europe, several genomic studies based on fossil remains have provided us with information about the peopling of Asia and the relationship between the populations living on that continent. For example, the sequencing of the genome of an individual dating from 24,000 years ago, discovered on the Mal'ta site, near Lake Baikal, in the south of Siberia, has revealed surprisingly strong genetic affinities with the populations of western Eurasia and the indigenous people of the Americas, rather than with present-day East Asians or Siberians, which widens the gene pool of ancient Europeans to North Asia. Today we find genetic contributions of the ancient Mal'ta population at a level of 10% to 20% in Europeans and 30% to 40% in Native Americans. Subsequent studies have demonstrated that the Mal'ta divided about 15,000 years ago into two large groups: the hunter-gatherers of Europe to the west and the populations of North Asia to the east. There would seem to be a genetic continuity in North Asia from the Upper Paleolithic to the Bronze Age, about 5,000 years ago.

Replacement versus continuity

Starting in the Bronze Age, migration and admixture events that took place in Central Asia and Siberia seem to have led to the replacement and disappearance of this ancient component of North Asia through migrations from the west. These may have been migrations associated with the Yamnaya culture, which also moved towards Europe at the same time, and migrations from the east, like those linked with the Neolithic culture in Asia. Thus, the populations to which the Mal'ta may have belonged, as well as the Ust'-Ishim Man, also found in Siberia and dating back 45,000 years, have left few visible genetic traces in modern Eurasians. These facts underline the importance of population replacements in the course of human history.

By way of contrast, it is also worth emphasizing the cases of genetic continuity: there are examples of groups of ancient individuals who contributed to the current genetic diversity of human populations. This is the case with Tianyuan Man, found in Tianyuan Cave in China. He is 40,000 years old, and his genome was sequenced in 2017. His genetic make-up is closer to that of East and South-East

Asians than to that of Europeans, both ancient and current, revealing a genetic continuity in East Asia over the last 40,000 years. Another interesting observation is that the genome of Tianyuan Man is close to the genetic diversity observed today in some Native American populations from South America, suggesting that the settlement of the Americas was carried out by different ancestral Asian populations, themselves already differentiated, although this is still debated. Unlike the European continent, where these events are rare, we find in Asia other examples of population continuity. This is illustrated by the sequencing of the genomes of individuals who lived in Tibet, North Korea or Japan about 7,000–10,000 years ago: they prove to be very close to the populations living in these regions today.

The history of East and South-East Asia

Several studies of the ancient DNA of individuals who lived in East and South-East Asia have provided us with valuable information about the settlement of these regions. For example, they have shown that, at the beginning of the Neolithic, the populations of what is now North and South China were more clearly differentiated genetically than they are today. This has been interpreted as evidence that, during the Neolithic, populations from the north, who probably grew millet, migrated towards the south of China, where rice was cultivated, and admixed with the inhabitants of that region. The genetic data also suggest that the populations speaking Austronesian languages, which are found today from Taiwan to Polynesia, came originally from the south-east coast of continental East Asia, given their genetic similarities to the fossil remains dating from the Neolithic found in that region.

As for South-East Asia, the paleogenomic data reveal that the modern populations of this region are the result of the admixture of several ancestral populations: the continental Hoabinhians, the first inhabitants of the region, related to the present-day populations of the Andaman Islands (Onge, Jarawa) and to the Jahai,

71

hunter-gatherers of Malaysia; the populations originally from what is now southern China, who introduced agriculture (especially rice cultivation) into the Asian Peninsula about 4,000 years ago; populations speaking Austro-Asiatic languages, descended from the admixture in the Neolithic between the Hoabinhians and the first farmers from the north; and populations speaking Kra-Dai and Austronesian languages, whose expansion may have introduced a new genetic ancestry from East Asia into continental and island South-East Asia, respectively, about 2,000 years ago.

The complexity of Central Asia

One part of Asia has been relatively well studied, from the point of view of both modern and ancient DNA: this is Central Asia, broadly speaking – the vast region bordered on the west by the Caspian Sea, on the east by Mongolia and China, on the south by Iran and Afghanistan, and on the north by Russia. This region includes two geographical areas often studied separately but that in fact share a common history and culture: on the one hand Central Asia strictly speaking, corresponding to the republics of Kazakhstan, Kirghizstan, Tajikistan, Turkmenistan and Uzbekistan, and on the other hand North Asia, which includes western Mongolia, part of Russia – southern Siberia – and the westernmost region of China, the autonomous Uyghur republic of Xinjiang. This region also presents a rich variety of natural environments, from the steppes of Central Eurasia to the boreal forests (taigas) of the North and the tundra on the periphery of the Arctic Ocean.

A genetic study conducted in 2019 on populations living in the migratory corridor constituted by countries as diverse as Armenia, Georgia, Kazakhstan, Moldova, Mongolia, Russia, Tajikistan, Ukraine and Uzbekistan showed that these populations can be divided into three broad genetic ancestry groups associated with particular geographies and linguistic affiliations. As far as geography is concerned, these groups reflect the different ecoregions of Inner Asia, that is, the boreal forests and the tundra, the wooded steppes and the shrubby steppes, while, as regards languages, they correspond to the Uralic languages and the Turkic languages of the North and South. Some populations, however, deviate from this general trend, demonstrating the importance of recent admixture and later population movements, which wiped out the genetic continuity we would expect in some of these groups. For example, a study of individuals who lived in the steppes of Central Asia in the last 4,000 years has shown that different migratory events transformed the genetic landscape of the Eurasian steppes after the Bronze Age. Occupied at first by populations from the west, speaking Indo-European languages, these steppe populations gradually transitioned to today's largely Turkic-speaking groups, who are mainly of East Asian origin.

South Asia and the Indian gradient

Ancient DNA data from South Asia is unfortunately rare: the specific climatic conditions of this region have led to the poor preservation of DNA in fossil remains. Nevertheless, the peopling of the Indian subcontinent has been studied in detail. What emerges is that, surprisingly, the modus operandi of its peopling resembles that of Europe in its tripartite composition. Today's Indian populations are the result of the admixture of three genetic components attributed to three ancestral peoples. South Asia was initially inhabited by groups of hunter-gatherers who settled there about 50,000 years ago, before the arrival of agriculture and just after the exodus of *Homo sapiens* from Africa. Subsequently, episodes of admixture between the first farmers from the Zagros Mountains (in present-day Iran) and groups of hunter-gatherers, both local and coming from western Siberia, led to the formation of a new population, which gave rise to the Indus Valley civilization some 3,700–5,400 years ago. After the decline of the Indus civilization about 3,500 years ago, this population, already the result of heavy genetic admixture, admixed even more: on the one hand with local hunter-gatherers, giving rise to the ancestral population of southern India speaking Dravidian languages, and on the other hand with peoples from the Eurasian steppe speaking Indo-European languages, constituting the ancestral population of northern India.

The proportions of these two ancestral genetic components vary in today's India in the form of a north–south gradient, the *Indian gradient*, with larger ancestry proportions of steppe origin in the north and ancestry connected rather to the local hunter-gatherers in the south. The peoples who carry the highest frequencies of the northern ancestral component are those who speak Indo-European languages and belong to the highest castes, like the Brahmins, while the southern ancestral component is mostly represented in peoples speaking Dravidian languages in the south of India, and in the Onge hunter-gatherers of the Andaman Islands – the closest living representatives of this ancient genetic component.

Oceania: the last frontier

The long journey of human migrations does not stop in continental Asia. It continues rapidly, reaching the islands of Oceania. From a geographical and cultural point of view, the South Pacific was historically divided into several regions that owe their names to the French explorer Jules Dumont d'Urville, who traversed the Pacific during two voyages in 1820. He coined the names Melanesia ('black islands'), Micronesia ('small islands') and Polynesia ('many islands') – the definitions are imprecise, but the names have remained. A more relevant division of the Pacific can be obtained by examining the chronology of the region's peopling, thereby separating Near Oceania, populated more than 40,000 years ago, from Remote Oceania, which was not occupied until the last 3,000 years.

The settlement of Oceania comprises two extremes: the end point of the first human migration out of Africa, it is also the arrival point of the last migrations towards territories hitherto unoccupied by humans – we did not set foot on New Zealand soil until just 730 years ago. Indeed, according to the archeological data, Sahul experienced human occupation between 40,000 and 60,000 years ago. As for the Bismarck Archipelago and the Solomon Islands, which today form Near Oceania along with New Guinea, we find traces of occupation there between 30,000 and 40,000 years ago. The descendants of these first settlers are Papuan-speaking communities living today in New Guinea and on some islands of the Bismarck and Solomon archipelagos. In the Pleistocene, humans went no further than Australia and Near Oceania: it would be about another 35,000 years before they reached the rest of the Pacific.

Australians and Papuans

Whole-genome data indicate that today's populations of Aboriginal Australians and Papuans are descended from a single population, originally from Africa, which diverged from the Eurasian populations about 60,000 years ago. Australians and Papuans subsequently separated 25,000–40,000 years ago, testifying to an ancient structure and genetic isolation preceding the rise in sea level and the submerging of Sahul 10,000 years ago. The genetic data also show that the different groups of Aboriginal Australians are highly differentiated among themselves, and that their separation into several distinct groups dates from the last 10,000–32,000 years, a period that coincides with environmental changes linked with desertification.

Sailing without map or compass in the South Seas

There is one remarkable fact about the peopling of the Pacific, which is that it involved the first great sea crossings in human history. Even though the sea level in the Pleistocene was more than 100 metres lower than it is today, Sahul was not connected to the Asian continental mass (called Sunda), which at the time encompassed what are today the islands of Sumatra, Java and Borneo. But the humans of that time did not set off blindly: they could see lands on the horizon. In fact, it is likely that, for those who attempted these crossings, the lands they hoped to reach would mostly have been visible from about 70 kilometres. By contrast, the peopling of the rest of the Pacific – which is called Remote Oceania, and which includes the islands of Santa Cruz, Vanuatu, New Caledonia, Fiji, Micronesia and Polynesia – probably involved crossings ranging from at least 400 kilometres to – in the case of some of the most remote islands – thousands of kilometres on the high seas. This was a highly adventurous undertaking, and one impossible without much more sophisticated navigation techniques.

Unlike Near Oceania, Remote Oceania was only populated in the last 3,500 years – the most recent expansion of humans into large empty lands. This expansion began some 5,000 years ago, probably in Taiwan according to the linguistic data, and continued across the Philippines and Indonesia, reaching the already inhabited lands of Near Oceania and finally reaching Remote Oceania. From the linguistic and archeological point of view, the settlement of Remote Oceania was associated with the expansion of the Austronesian languages, the development of rice cultivation and the Lapita cultural complex, characterized by the production of certain kinds of pottery that appeared for the first time in the Bismarck Archipelago about 3,300 years ago. About 400 years later, the Lapita culture reached the islands of Tonga and Samoa in the western part of the Polynesian triangle formed by Hawaii to the north, New Zealand (Aotearoa) to the south-west and Easter Island (Rapa Nui) to the south-east. Central and Eastern Polynesia would be settled in a later wave, separated from the first by about 1,500 years. The first known dates for the peopling of Central Polynesia are around 1,200 years ago for the Society Islands, while the last lands in the world to have been settled are Easter Island and New Zealand about 730 years ago.

A complex puzzle of migration and admixture

Genetically speaking, the populations living today in the South Pacific are, for the most part, the result of the admixture between two ancestry components associated with the two migration waves of the Pacific – the 'Papuan' component and the 'Austronesian' component. The proportions of these genetic components vary considerably from one population to another: while in Near Oceania the Austronesian component is about 20%, in Remote Oceania and especially in Polynesia it can be as high as 80%. Nevertheless, the genomic data derived both from modern populations and from ancient DNA have also revealed that the South Pacific was an important corridor for human migrations and that its settlement is much more complex than the two waves of migrations might have suggested.

Studies conducted in 2018 by Mark Stoneking's team, from the Max Planck Institute in Leipzig, analysed the genetic profiles of more than fifty populations of Near and Remote Oceania. They showed that the dispersal of populations from Near Oceania to Remote Oceania unfolded in a leapfrog fashion, for example completely bypassing the principal chain of the Solomon Islands. Thus, some populations show a much larger than expected Papuan component – the inhabitants of the Santa Cruz Islands, for example. This suggests that there were different waves of migration, separated in time, that affected the islands of the Pacific in different ways.

The complexity of the peopling of Oceania is clearly supported by ancient DNA data from the islands of Vanuatu and Tonga. The genomes of individuals who lived at different periods from the beginning of the Lapita period, 3,000 years ago, to the present day have been sequenced. What emerges is that individuals from the Lapita period were almost exclusively of Asian origin and close to present-day Taiwanese, which is only to be expected if we consider the hypothesis that the Austronesian expansion came from Taiwan. Nevertheless, most of the other individuals who lived between 2,300 years ago and today share at least 70% of their genome with modern Papuans! These observations indicate that there was another migration wave, likely of people of Papuan ancestry, some centuries after the first settlement of Vanuatu, which largely replaced the original Lapita populations.

The maze of Oceanian genomes

In my laboratory, we recently investigated the demographic history of Oceania by sequencing the genomes derived from twenty populations of Near and Remote Oceania. Using the most advanced methods for reconstructing the demographic past of populations, we first noted that the ancestors of the peoples of Near Oceania, such as the populations of Papua New Guinea, the Bismarck Archipelago and the Solomon Islands, experienced a severe bottleneck just before the settlement of the region and separated about 20,000–40,000 years ago. We also showed that the Austronesian expansion – thought to have begun in Taiwan about 5,000 years ago – was not followed by an immediate admixture event with peoples of Near Oceania, but rather involved different waves of migration and admixture with the local Papuan populations. In addition, the data, which are based on the DNA of present-day populations, complement those derived from ancient DNA and shed new light on the remarkable complexity of the peopling history of the Vanuatu Archipelago. We established in fact that the 'Lapita' Asian component present today in Vanuatu was also brought later by peoples of Papuan origin, who had already admixed with Austronesian peoples. Overall, what emerges from this genomic study is a scenario in which very complex and repeated population movements occurred in the South Pacific.

The Kon-Tiki: *exchanges between America and Polynesia?*

The complexity of the peopling of the Pacific reached its climax in Polynesia. The rare genetic studies of Polynesians show that most of their genetic make-up is of Asian origin (about 80%), close to that of the Taiwanese or the Filipinos. The rest is of Papuan origin, in variable proportions depending on the islands, particularly in Western Polynesia, or of European origin, mainly in the Marquesas or on Easter Island. In the history of the region, one question remained very controversial: were there early contacts – before the

78

European arrival – between Polynesian and Native American populations? This hypothesis presupposes a direct link by sea between Polynesia and South America, which would have been a formidable leap into the unknown. The idea, which has long sparked debate, was initially put forward by the Norwegian explorer Thor Heyerdahl in 1947. To prove that such a voyage was quite possible, Heyerdahl himself undertook the crossing from Peru to Polynesia on board a wooden raft, the famous *Kon-Tiki*. His achievement did not put an end to the controversy. Several studies have examined the subject by using genetic data derived from both ancient DNA and modern populations. But they reached opposing conclusions . . .

Finally, in 2020, the question seems to have been settled. Researchers from the universities of Stanford in California and Irapuato in Mexico conducted a study of more than 800 individuals – Native Americans and inhabitants of different islands of the South Pacific. Their work establishes that there was indeed admixture between Native Americans from South America, the origins of whom are close to the present-day populations of Colombia or Ecuador, and Eastern Polynesians, and that this admixture took place about 800 to 900 years ago. In other words, this exchange occurred well before the arrival of Europeans in Polynesia about 200 or 300 years ago. According to the authors, the likely site of this encounter was the island of Fatu Hiva, the southernmost of the Marquesas – exactly as Thor Heyerdahl predicted more than seventy years ago. So, although Easter Island is the closest to the South American coast, it is not the site of the first contact, since this contact took place *before* the peopling of Easter Island. How then did these Native American peoples, confronting the high seas, get as far as the Pacific islands, more than 7,000 kilometres from the American coast? The authors of the study suggest the hypothesis of drifting at sea: at these equatorial latitudes, a boat coming from South America would be pushed from east to west by the winds and by strong sea currents, in the direction of the Marquesas Islands.

Denisovan territories

The Oceanian peoples, being behind the greatest maritime epic of all times, have richly deserved their reputation as great travellers. But they have other remarkable features. Their genomes have the highest proportion in the world of so-called 'archaic' material, in other words material from other human species: up to 6% in some Pacific populations. As with the rest of the non-African populations, they have inherited about 2% of Neanderthal material, testament to an episode of ancient admixture 50,000–55,000 years ago, just after the exodus from Africa of *Homo sapiens*. Moreover, Oceanians, and in particular the populations of Melanesian origin, also present the highest proportion of genetic material of Denisovan origin in the world.

In the South Pacific, Denisovan ancestry may in fact reach as much as 3% or 4% in the Papuans, but also in populations having a strong genetic affinity with the Papuans, such as the inhabitants of the Bismarck Archipelago, the Solomon Islands or the Vanuatu. We also find populations of hunter-gatherers in the Philippines, the Negritos, who present about the same proportion of Denisovan material in their genomes, followed by the populations of East and South Asia, who only present a few genomic traces of this ancient hominin, at a level of 1% at most. Similarly, populations having Asian ancestry, like the Native Americans, South Asians, or those who have received gene flow from Asians, like the Finns in Europe, may also present Denisovan genomic traces, but at a proportion of less than 1%.

All the same, it is surprising that the genetic epicentre of the distribution of Denisovan vestiges in today's populations is located in the South Pacific – in other words, more than 8,000 kilometres from the Siberian cave where the only representative of the Denisovans was found, admittedly reduced to a piece of a phalanx, but yet still available for genome sequencing. This raises many questions about the distribution area of these archaic hominins.

A study based on archeological data, carried out by Jean-Jacques Hublin's team, has made it possible to enlarge the distribution area of the Denisovans to the Tibetan plateau, thanks to the analysis of ancient proteins from a mandible dating from more than 160,000 years ago.

From the genetic point of view, even though we only have a single genome that really pins down the Denisovans, we can retrace their distribution area through the study of the Denisovan genetic segments present in modern populations, and thus gain a better understanding of the history of these Asian hominins. Several genomic studies suggest that they may have lived up to the borders of the Wallace Line, which separates, on the one side, what are now the islands of Sumatra, Java and Borneo – at the time, part of the Asian continental mass known as Sunda – and, on the other side, the archipelago of the Philippines, the Indonesian island of Sulawesi and the ancient Sahul.

The Denisovans were probably not the only archaic hominins to have crossed the Wallace Line: as attested by the presence of Callao Man (*Homo luzonensis*) in the Philippines at least 50,000 years ago, and Flores Man (*Homo floresiensis*), who lived on the Indonesian island of Flores up to at least 50,000–60,000 years ago. This would make the Asia-Pacific region the scene of a – probably fruitful – cohabitation between the first modern humans who arrived in the region and other, now-extinct hominins.

Recurrent archaic admixture

Whereas admixture with the Neanderthals seems to have occurred in one go in the ancestors of the Eurasians, the most recent genomic data show that there were several admixture events between modern humans and the Denisovans, at different times and between different groups. Studying the genetic variability of the Denisovan segments in today's populations may teach us about the genetic diversity of the Denisovans themselves, as well as about their levels of population structure. Comparison of these Denisovan segments present in modern populations of the Asia-Pacific region with the Denisovan genome from the Altai has shown that the Denisovans formed indeed very heterogeneous groups. Some of these groups are genetically close to the Altai Denisovan specimen, whereas others are relatively distant from it.

According to the different studies that have tackled this question, there were at least three independent admixture events between Denisovans and Sapiens. The Denisovan population closest to the Altai genome was the one that contributed genetic material to East Asians and Siberians. On the other hand, two different Denisovan populations – genetically distant from the Altai specimen – were those that admixed with populations of Papuan origin. These two Denisovan populations separated from the Altai Denisovan between 200,000 and 400,000 years ago, which raises questions as to whether they truly belonged to the 'Denisovan' lineage. Indeed, their separation from the Altai Denisovan is close chronologically to the divergence between Denisovan and Neanderthal, which dates from about 500,000 years ago. This suggests that these Denisovan groups may represent an independent archaic form. Several studies, including those of my laboratory, even support a fourth admixture event between the Denisovans and the Negrito hunter-gatherers of the Philippines.

These genomic studies have also made it possible to date the periods during which these ancient admixture events may have taken place. The timings point to a period stretching from about 20,000 to 50,000 years ago, which indicates that the Denisovans survived the Neanderthals, who died out about 40,000 years ago. A study conducted by Murray Cox's team from Massey University in New Zealand, on the basis of an admixture event with the Papuans dating back just 14,500 years, has even advanced the hypothesis that the Denisovans may have survived until the very end of the Pleistocene. If these estimates are confirmed, they would make the Denisovans the last non-Sapiens hominins to have died out.

Off to conquer the Americas: the crossing of the Bering Strait

The human history of the Americas is also one of a kind. The peopling of these two continents represents the high point of the last Late Pleistocene expansion of anatomically modern humans who had left Africa. In addition, it involves the encounter between, on the one hand, Native Americans, the native peoples of America, originally from Asia, and, on the other hand, from the fifteenth century onwards, peoples from Europe and Africa. For this reason, the genetic diversity of today's American populations could be said to represent the genetic diversity of a large part of the world. Different factors have led to this present-day diversity: the variable density of the Native population on the arrival of the European colonizers, the extent of more recent European immigration to specific geographical areas, the degree of involvement of some parts of the New World in the slave trade from Africa, and the different levels of admixture between Native Americans, Europeans and Africans.

The Clovis culture

To begin with the initial settlement of the American continent: according to archeological data, the human presence in America dates back at least 14,500 years and is associated with the Clovis cultural complex, named after the town of Clovis in New Mexico, where the first vestiges of this culture were discovered in the late 1930s. The Clovis culture, characterized by fluted projectile points, emerged 13,500 years ago. Its traces are clearly observed in North America. The study of this cultural complex confirmed that the first inhabitants of the American continent had come from Asia via the Bering Strait, discrediting other – occasionally somewhat preposterous – hypotheses, such as the so-called 'Solutrean' hypothesis that the first Native Americans were of European origin, or the idea that they were one of the ten lost tribes of Israel. For a long time, the scientific community agreed that the peoples associated with the Clovis culture represented the first inhabitants of America and that their arrival led to the extinction of the American megafauna: mammoths, giant sloths, mastodons and so on.

Over the last few decades, however, many archeological discoveries have, through their date estimations, challenged the theory that bearers of the Clovis culture were the first settlers in America. On the basis of human coprolites, for example, a human presence not associated with the Clovis culture has been noted in the Paisley Caves in Oregon, 14,000 years ago. At the Meadowcroft site in Pennsylvania, we also find blades and lithic cores dating back 16,000 to 19,000 years and which consequently precede the Clovis culture – although these dates remain controversial. In the south of Chile, at the Monte Verde site, we find stone tools as well as organic remains, like charcoal, that also precede the Clovis culture. Today, a pre-Clovis human presence in America is accepted on the basis of the archeological data. If the question of the culture or cultures that gave birth to modern Native Americans remains without a clear answer, genetic studies, especially those derived from ancient DNA, have allowed us to progress somewhat in our knowledge of the subject.

Linguistic families and migrations

Before tackling the genetic contribution to this topic, we should also examine the hypotheses derived from linguistic studies on the peopling of America. In 1986, the linguist Joseph Greenberg put forward a hypothesis that, although controversial, caused a great stir in the scientific community and triggered a large number of genetic studies setting out to test it. Greenberg's theory postulates that America was peopled in three great waves of migration corresponding to the three linguistic families observed today: the Eskimo-Aleut languages spoken in the far north of the continent, Greenland and some parts of Siberia; the Na-Dene languages spoken in western Canada, Alaska and the south-western territories of the United States; and the Amerindian languages, which encompass the remaining Native American languages.

Once again, it is genetics that has made it possible to throw light on the question of the number and nature of the migrations that led to the peopling of the Americas. The studies tend to support the hypothesis of at least three independent waves of dispersal. Most Native American populations share great genetic similarities that differentiate them from the rest of the world, and they derive from a single source of Beringian origin. This would mean that the majority of Native Americans are descended from a single migratory event that took place about 15,000 years ago, whereas the Eskimo-Aleut- and Na-Dene-speaking populations derive their genetic ancestry not solely from two other independent migratory events, as proposed by Greenberg, but rather from a mixture of different waves of migration, including the first great migration and at least two other, more recent expansions.

The genomic data have also refuted certain hypotheses, based on cranial morphology and lithic analysis, which suggested that the first Americans were of European, Polynesian or even Japanese origin. In 2014, the genome of an individual dubbed Anzick-1, who lived 12,600 years ago in what is now Montana and is associated with the Clovis cultural complex, was analysed. The sequencing showed that the population to which this individual belonged was indeed ancestral to most current Native Americans, which points to a genetic continuity in America dating back at least 13,000 years. One interesting point should be underlined: the genetic variability of this individual is more closely linked to the present-day Native populations of Central and South America than to certain groups in North America, which suggests an ancient population structure in America dating back at least 12,600 years. Another clue appeared in 2015, with the sequencing of the genome of Kennewick Man, who lived in Washington State about 9,500 years ago. The genetic diversity of this ancient individual proved to have close links with present-day Native Americans. This new element strengthens the idea that most Native Americans share a common origin.

Four stages

Studies in the last few years have also shed light on the processes leading to the first settlement of America. What emerges is that this peopling occurred in four large phases:

- First phase: separation of the ancestors of the Native American and the Asian peoples.
- Second phase: prolonged isolation (near Beringia) of the population ancestral to the Native Americans.
- Third phase: arrival (or arrivals) of the paleo-American peoples on the continent and spatial expansion.
- Fourth phase: different admixture events, in historical times, of the Native American peoples with populations of European and African ancestry.

The ancestral population of the Native Americans probably began to diverge from the East Asians about 36,000 years ago, somewhere in North-East Asia, and subsequently admixed for several thousand years with two other populations: a population of ancestors of the Han Chinese and a population close to Mal'ta Man. As we have seen, the genome of Mal'ta Man, who lived about 24,000 years ago in southern Siberia, presents genetic affinities not only with the populations of Western Eurasia but also with today's Native Americans, who carry traces of the Mal'ta population to levels of between 30% and 40%.

The second phase corresponds to a long period of maturation during which this admixed population, the ancestors of the Native Americans, remained isolated until about 16,000 years ago, probably in Beringia. It was at the end of this period that the third phase began, that is, the expansion of the proto-American population into the continent. This expansion seems to be concurrent with a divergence of the population into two branches, dated to 4,600–17,500 years. One branch, which led to the Native Americans of the north, is associated with the genetic diversity of Kennewick Man, while the other branch gave birth to the Native Americans of the centre and the south and is associated with the genetic diversity of the previously encountered Clovis specimen Anzick-1. The geographical location where this divergence took place remains unclear, but it is likely that it corresponds to an area situated between eastern Beringia and the non-frozen region of North America. Following this first arrival, the peopling of America was exceptionally rapid: in less than 1,500 years, it reached as far as Tierra del Fuego, in the far south of the South American continent. As for the routes followed by this first great migratory wave, two hypotheses have been advanced: one following the coasts of the Pacific and the other, inland and more circumscribed, following an unfrozen corridor of about 1,500 kilometres.

So most of the modern Native American populations are descended from a single population that entered the continent about 16,000 years ago. However, at least two other migration waves took place in the Arctic regions. One is at the root of the Paleo-Eskimos about 4,500 years ago: it is associated with the Saqqaq and Dorset cultures. The other gave birth to the Neo-Eskimos just over 1,000 years ago, linked to the Thule culture. The sequencing of ancient genomes derived from Paleo- and Neo-Eskimos have contributed in no small measure to a greater understanding of the history of the peopling of the American Arctic. The first ancient human genome to have been sequenced is that of an individual from the Saqqaq culture who lived in Greenland 4,000 years ago. This sequencing revealed that the population to which this individual belonged was genetically close, not to the present-day Native Americans and Inuit, but to the Koryaks and Chukchi of Siberia, which supports the hypothesis of an independent wave of migration. Genomic analysis of further specimens from the Dorset and Thule cultures has confirmed the distinct origins of the Paleo-Eskimos, who were replaced over the following 700 years by another wave of migration associated with the Thule culture, the ancestors of the modern Inuit.

Finally, a fourth wave of migration has been suggested, independently, by the teams of David Reich at Harvard and Eske Willerslev in Copenhagen, on the basis of the discovery of a genetic signature of 'Australasian' origin, the so-called 'Ypikuéra population' or 'Y population', presenting genetic characteristics close to the peoples of Australia, Papua New Guinea and South-East Asia, in some Native American groups in Brazil. However, it has not yet been established if this presence of the Y population in America is a methodological artefact, the result of more recent genetic exchanges, or if it attests to the existence of a true population contributing additional ancestry to the Native Americans.

The slave trade and the colonial period

The peopling of America does not stop with the migratory history of the ancestors of Native Americans. It also involved, from the late fifteenth century onwards, the encounter between millions of Native Americans and populations of European and African origin. Three genetic components saw admixture events of varying intensity, under the effect of demographic and sociocultural factors, which had a major impact on the genetic diversity of modern populations across the continent. It is with the arrival of Christopher Columbus in the Bahamas in 1492 that European immigration began. The first arrivals were immigrants of Spanish and Portuguese origin, among whom men were over-represented, and who settled in different regions of the Caribbean, the Pacific coasts, and in certain regions of the interior already inhabited by Native Americans. In the century that followed, other Europeans of English and French origin settled in the Caribbean and along the Atlantic coast, mainly in North America. As for the arrival of the component of African origin, it began early in the colonial period with the Spanish and Portuguese slave trade, and continued with the British and French powers, particularly in the context of the exploitation of sugar plantations. The slave trade became more widespread with the collapse of the indigenous population: it is estimated that about 90% of Native Americans perished following the arrival of the Europeans.

During the colonial period, the arrival of Europeans and Africans had a significant impact on the history of America, since about 2 million Europeans and 10 million Africans arrived at this time. Even though the abolition of slavery during the nineteenth century slowed the large-scale arrival of Africans, European immigration continued until the beginning of the twentieth century, to both North and South America. It is important to underline that relations between European colonists and populations of Native American and African ancestry were very different depending on whether we consider the British colonists or those of Iberian origin. Throughout this period, there were no doubt much stronger social barriers between Europeans and non-Europeans in British America than in Latin America, which had a considerable impact on the degree of admixture between populations of different origins.

The admixture resulting from the different social and cultural relations between the peoples that have inhabited America has left significant traces on the genetic diversity of modern populations. In its turn, the study of this genetic diversity allows us to dissect the demographic past of America as a whole and infer the relative contribution of the Native American, European and African components when there is a lack of historical data. If we exclude those populations who define themselves as being of exclusively Native American origin, the genetic study of current populations across the American continent clearly recapitulates the colonial history of this continent. Thus, the populations of North America (Canada and United States) are predominantly of European origin (more than 85%), apart from Greenland, whose population is 75% of Native (Inuit) origin.

In Central and South America, on the other hand, the genetic data illustrate the major importance of admixture in the history of these regions, confirming the way in which the flexibility of social barriers favoured admixture. The two most represented genetic components in Latin America are therefore that of European origin, which varies from 20% to 60%, depending on the populations, and that of Native American origin, varying from 20% to 80%. For example, European genetic ancestry is dominant in Costa Rica, Argentina and Brazil, where it accounts for more than 50% of the genetic make-up, while the Native American component is particularly high in Bolivia, Peru, Mexico and Guatemala, where it reaches 60% to 80%.

As for the African component, it is present in low frequencies, about 5% to 10%, in most of the populations of North, Central and South America that do not self-identify as of African origin, except for Brazil and Colombia, where it reaches 20%. As may be expected, this African component is much better represented within populations that consider themselves as being of African origin, as is the case with African Americans in North America, where it is present at a level of about 80%. Among the communities that self-identify as of African origin in Latin America, on the other hand, the African genetic component reaches 30% to 50%, illustrating the fact that the social barriers preventing admixture were weaker in Latin America.

Colonial disaster in the Caribbean

The situation of the Caribbean appears completely different from that of continental America. In fact, the largest genetic component in the general population is of African origin – it reaches 80% in most of the islands, except for Cuba and Puerto Rico (about 20%) – followed by the European component. The most striking thing about the genetic composition of Caribbeans is the extremely low presence of the Native American genetic component. It is found in a proportion of less than 10% in all populations, except in the Dominican Republic and Puerto Rico, where it is as high as 20%. The scarcity of the Native American genetic component in the Caribbean shines a light on the consequences of the arrival of the European colonists and the populations that emerged from the slave trade: the almost total annihilation of the Native populations of the region.

A long history of admixture

This overview shows that we are the result of a long history of admixture and genetic exchanges. The genetic studies are categorical. No clear break or barrier separates the different human populations: they constitute a continuum of genetic diversity, higher in Africa and decreasing as we get further away from the continent that was the cradle of humanity. What emerges from the analysis of the diversity of our genomes is that there are no 'pure' populations, since all human beings, independently of their geographical or ethnic origins, are the result of a long history of admixture. Each of our genomes is in reality a true mosaic of DNA segments deriving from the different ancestral populations that have admixed over time – from our origins to the present day.

The earliest admixture event documented so far occurred as soon as humans left Africa and met the Neanderthals, somewhere in Eurasia. Comparing the genetic diversity of modern populations with the genomes of Neanderthals has allowed us to answer a long-standing question in anthropology: was there hybridization between archaic humans and modern humans? Much to everyone's surprise, the genomic data have shown that there was indeed admixture between the two groups. Other human forms have contributed to making us what we are. When humans arrived in the Asia-Pacific region, about 50,000 to 60,000 years ago, they encountered the Denisovans, with whom they admixed on multiple occasions and in different regions. Another major surprise is that the last admixture events between modern humans and Denisovans have been dated to only about 20,000 to 25,000 years ago.

We know today that *what we are* is not only the product of several admixture events with archaic humans, but also, as we shall see in Chapter 5 of this book, that these archaic humans brought us genetic variants that were beneficial to our survival, especially variants associated with adaptation to cold or resistance to infectious agents, particularly viruses. If archaic humans were so well adapted to their environment, to the extent that they even contributed to our Sapiens ancestors being better adapted, why did they die out? For a long time, several theories based on geology and paleoanthropology circulated, imagining various reasons for their disappearance: climatic changes, volcanic eruptions, diseases of all kinds. The reality is probably much simpler and less flattering to our ancestors. It was probably the spread of *Homo sapiens* that brought about the disappearance of their archaic cousins.

The most likely explanation is that our Sapiens ancestors shared the same ecological niche as archaic humans and one group had greater reproductive success than the other and so has survived to the present day. The question is therefore not to find a reason for the disappearance of archaic humans, but rather to understand what the factors are that make us such an 'invasive' species.

In fact, the one thing that the different extinctions of archaic humans across the planet have in common is precisely the arrival of *Homo sapiens*. Whatever the geographical context, all the groups of hominins with whom Sapiens cohabited – *Homo neanderthalensis* in Eurasia, *Homo denisovensis*, *Homo floresiensis* and *Homo luzonensis* in East and South-East Asia, and *Homo naledi* in Africa – disappeared after the arrival of *Homo sapiens*. As Jean-Jacques Hublin has explained, because of the way in which our species exploits the environment and interacts with others, the story always ended the same way: the other human groups were replaced and absorbed by *Homo sapiens*. Other researchers see things more optimistically and assert that the Neanderthals, like the Denisovans, never disappeared: they continue to exist within us.

3.

Adaptation and Environment

What are we, when it comes down to it? Part of the answer lies in our history of settlement and admixture. But genetics allows us to say more: we are also the result of more than 200,000 years of biological adaptation to our environment. Adaptation is the mechanism that allows the species to change when the environment imposes constraints on organisms, constraints that 'select' those who are lucky enough to be endowed with characteristics that give them a comparative advantage. In this chapter, I will examine the way in which our species has adapted to its environment, presenting the different forms that natural selection may take and showing some of its effects, which can be 'read' in our genomes as we gradually learn to decipher them: language skills, skin colour, height, ability to digest milk, and resistance to cold and to a lack of oxygen, among many others. This history is enlightening, showing as it does the astonishing inventiveness of human life.

The vast spectrum of phenotypic diversity that can be observed in human populations today – our physical appearance, our different ways of metabolizing certain foods, our different reactions to diseases – is, to a large extent, the result of natural selection. Studies aiming to gain a better understanding of the part played by natural selection in the adaptation of human populations to their environment can benefit from the tools of genomics, which make it possible for example to detect signs of selection in certain genes, and even to differentiate them depending on the different forms that natural selection can take. What follows is a panorama of the results obtained using genomic approaches.

The challenges of adaptation

Humans left Africa and peopled the planet in less than 60,000 years. Today, there are humans on every territory above sea level in the world, and in the most varied environments: from the hot, dry savannahs of Africa to the sunless, inhospitable colds of the Arctic, from the rainforests of Central Africa or South-East Asia to the extreme and hostile conditions of life at high altitudes, in the Himalayas or the Andes, where oxygen is limited, by way of the concrete landscapes of New York or Paris, which sometimes take on the dimensions of gigantic megalopolises like Tokyo, Delhi, Shanghai or São Paulo.

This omnipresence of our species is worth pondering. Humans have encountered extremely different climates, they have had to deal with varied nutritional resources – sometimes scarce, sometimes abundant, sometimes hard to access, sometimes less so – and they have had to confront a very wide range of predators and pathogens. How have they managed to adapt? At what speed? By what mechanisms? Why is our physical appearance or our vulnerability to certain diseases so variable despite our genetic similarities and the relative youth of our species? The fact is that the world of phenotypic variation we observe in humans is vast. The differences between individuals are countless, often without consequences but occasionally associated with diseases. Although the environment plays an important role, genetic variation is also a major determinant of phenotypic variation. In this context, population genetics, and in particular the modern synthesis that resulted from the work of Ronald Fisher, Sewall Wright, J. B. S. Haldane, Theodosius Dobzhansky, Julian Huxley and Ernst Mayr, among others, have provided the theoretical and quantitative framework necessary to delineate the evolutionary determinants of the level of genetic, and therefore phenotypic, variability. In the course of their migrations across the globe, humans have been faced with diverse conditions to which they have adapted partly thanks to natural selection. This premise is the basis of the Darwinian theory of evolution – unlike Motoo Kimura's neutral theory, which, as we have seen, adopts a different position on this point.

The reality lies somewhere between the two. The Japanese scientist Tomoko Ohta, a disciple of Kimura, has suggested that most mutations are slightly deleterious, rather than neutral – a theory known as the 'nearly neutral' theory of evolution. These theories are not opposed in all respects: they all postulate that only a minority of the mutations present within a species, or between different species, are of an adaptive nature. With the progress made in genomics in the past few years, dissecting the molecular traces of the selection that has acted on the human genome, even when it is minimal, has proved crucial in identifying the genes that have led to the morphological and physiological diversity observed in humans. It has also led to a better understanding of the genetic architecture of adaptive phenotypes.

Natural selection: principles and forms

Natural selection relies on three great principles: the principle of variation, the principle of adaptation and the principle of heredity.

First and foremost, there has to be variation: diversity is the bedrock of natural selection. Without it, natural selection has no raw material to play with. In fact, it is the differences that may give rise to advantages – or disadvantages – for the organisms that carry them in a given environment: what we call selection is nothing but the translation of this relative advantage in terms of reproductive success.

Take the classic example of the length of a giraffe's neck. If all the giraffes on the savannah have necks of the same length, they all have the same skills – and the same limitations – in reaching the branches of the trees and eating the leaves. Evolution, as it has been understood since Darwin, supposes that there is variation. The principle of variation means that not all giraffes have necks of the same length: there are differences. Because of these differences, the giraffes with slightly longer necks are able to eat a little more, whereas, for other ones, the task proves more difficult. In competing for available resources, some therefore have an advantage. This is the second great principle, that of adaptation: giraffes with long necks are better fed than their peers with shorter necks, and so they are better adapted to their environment. The former have consequently a greater chance of surviving and reproducing: that is how we measure adaptation. Finally, the third great principle of natural selection is that of heredity: the giraffes with long necks transmit this 'trait' to their descendants and, since the latter are better adapted to their environment, necks will get longer in the giraffe population through Darwinian selection.

These principles are relatively simple. As so often, the reality of life ensures that they are actually much more complex. Natural selection operates in very varied forms, intervenes at different temporal and spatial levels, and its action presents different intensities.

At the level of the phenotype, to start with, we can distinguish three types of natural selection. First comes *stabilizing selection*, which favours moderate rather than extreme phenotypes, thus maintaining a status quo for a given phenotype. Take the example of height, which is a very variable quantitative trait in the human species. Northern Europeans are notably taller than southern Europeans, without there being any need to immediately compare them with other populations known for shorter stature, like some Native American groups, some populations in the Philippines, or the rainforest hunter-gatherers of Central Africa. In the case of *stabilizing selection*, both taller individuals and shorter ones would be disadvantaged compared with those of average height. Consequently, over time, most individuals in a population would be of moderate height. This regime diminishes the variance in height, but the average in the population remains unchanged.

The second type of natural selection is *directional selection*, which favours the prevalence of a specific extreme trait, like the short height of rainforest hunter-gatherers ensuring a greater chance of survival in the rainforests of Central Africa. With this type of natural selection, which is the basis of Darwinian evolution, the average height will decrease considerably in the population, even though the variance of the trait will remain largely unchanged.

Finally, a much rarer type of natural selection is *disruptive* or *diversifying selection*, which occurs when extreme phenotypes are advantaged compared with moderate ones, whose frequency decreases. In the example of human height, it is hard to imagine any reason why short or tall individuals should be advantaged by comparison with those of intermediate size. This is not the case with sticklebacks. For the length of these fish, we observe a selection of this type: the extreme sizes, both the shortest and the longest, have been favoured because of their adaptive effects – each of these extremes providing a particular advantage when it comes to accessing the available nutritional resources. Disruptive selection may potentially lead to a form of speciation, sympatric speciation, which occurs without geographical isolation.

As we have seen in the case of the giraffe's neck, natural selection works on the phenotypes. It is the characteristics presented by organisms (the longer neck), and which give them adaptive benefits or disadvantages, that are 'selected' and benefit from a reproductive advantage. But what is *inherited* is mainly the underlying genetics of the phenotype, which 'determine' it to a greater or lesser extent. So even if they are not the direct target, the genotypes are much influenced by selection. On the basis of genotypes, we can distinguish two forms of selection: when an allele is disadvantaged, we call this *negative* selection; if on the contrary it is favoured, this is *positive* selection.

Negative selection: purging deleterious mutations

Negative (or purifying) selection eliminates deleterious alleles from the population. It is the most widespread form of natural selection with regard to genomes. In humans, it is estimated that between 38% and 75% of all new mutations altering the amino acids are affected by moderate or strong negative selection. This frequent 'purge' of deleterious mutations may also lead to the accidental suppression of associated neutral mutations: this is known as *background selection*. The speed of the selective elimination of deleterious alleles depends basically on two factors: the deleterious character of the allele (or selection coefficient) and the effective size of the population (noted as *Ne*). A dominant lethal mutation will, by definition, be immediately eliminated in a single generation, while a mutation that is only slightly deleterious may maintain itself for longer in the population before decreasing in frequency. Similarly, in populations of large effective size (the effective population size being the theoretical number of ancestors in a population that explain current genetic diversity), natural selection is theoretically more effective, and the deleterious mutations will therefore be eliminated more rapidly than in populations with a small effective population size, where the effects of genetic drift are stronger and may in this way maintain these mutations.

Recent studies have deepened our understanding of the way in which the history of human populations in general, and their effect-ive population size in particular, may alter the efficacy of negative selection. Non-African populations have a higher proportion of deleterious variants in the homozygous state in essential genes compared with African populations. This observation is consistent with a less efficient negative selection within populations of smaller effective population size, like Europeans or Asians.

In fact, although the history of a population seems to have only a minor impact on the average burden of deleterious mutations on each individual, it has been reported that genetic drift has a greater impact on the frequency of slightly deleterious mutations in those populations resulting from a bottleneck compared with populations with a larger size. The number of deleterious homozygous geno-types carried by individuals increases in relation to their distance from Africa: the further we move from the African continent, the less genetic diversity we observe, but the greater the proportion of deleterious homozygous mutations. For example, both the hunter-gatherers and the farmers of Africa present a smaller number of deleterious mutations than Siberians or Mayas, two populations that present less genetic diversity, owing to the effects of the 'serial founder effects' that characterize, as we have seen, the process of the peopling of the planet by *Homo sapiens*.

This tendency is even more pronounced in populations that have undergone extreme demographic events that, for various reasons, have considerably reduced their effective population size. The Que-becois, resulting from a strong founder effect, or the Finns, who experienced a strong bottleneck, present an even greater proportion of deleterious mutations than populations that have experienced demographic expansions such as most European populations. In addition, these mutations may lead in some cases to a complete gene inactivation. A study, based on the sequencing of about 3,000 exomes (i.e. the part of the genome that encodes the proteins) from the population of Finland, has shown that this population presents a higher proportion of rare 'loss of function' mutations than other European populations. Some of these mutations are even associated with phenotypes of medical interest, such as the levels of vitamin B_{12} or the vulnerability to cardiovascular diseases.

The most marked effects of genetic drift in small populations, where natural selection has not yet had time to eliminate certain deleterious mutations, are clearly visible in the increased incidence of certain diseases within them. A few very striking examples are: congenital chloride diarrhoea, a rare, severe intestinal disease in children that particularly affects the Finnish population; familial dysautonomia, a severe alteration in the activity of the autonomic nervous system leading to multisystemic dysfunctions, which is present in some populations of Jewish origin; and hereditary deafness, observed particularly in some isolated communities in the central area of Costa Rica.

Positive selection: when humans start to speak

Unlike negative selection, which eliminates deleterious mutations, *positive selection* works on those mutations that are advantageous to survival. This mechanism operates on different timescales: we can observe its effects since our separation from our common ancestor with the chimpanzee 5 or 6 million years ago, or since the emergence of agriculture about 10,000 years ago. Positive selection may also lead to changes that are either shared by all the individuals of the same species and different compared with another species (*selection at species level*), or present in a specific human population (*local adaptation*). The study of natural selection at different periods can provide valuable information about specific evolutionary processes or critical periods of innovation.

A representative example of selection at species level is the *FOXP2* gene. In popular science, this gene is often called the 'language gene' – erroneously: it would obviously be naive to think that there is only one gene involved in a phenotype as complex as the articulated language characteristic of our species. In 2001, British researchers identified in the members of a single family a mutation that modified the protein sequence of *FOXP2*. The carriers of this mutation presented specific speech disorders: a difficulty in diction linked to deficiencies in the sequential movements of the mouth necessary to the articulation of words (verbal dyspraxia) as well as grammatical shortcomings.

The sequence of amino acids in the FOXP2 protein is highly conserved across vertebrates. We know that it regulates the development of various regions of the brain, including those that are involved in the production of speech, like Broca's area. Modern humans and Neanderthals share the same sequence of the protein, but this differs from the form present in chimpanzees, gorillas and macaques by two non-synonymous mutations. Thus, in the course of the last 6 million years of evolution, after the divergence between the ancestors of humans and the chimpanzees, two mutations accumulated only in the human lineage, while the sequence of this gene has remained practically intact during the 130 million years of evolution of the vertebrates, starting with the mouse.

It would seem, therefore, that an acceleration of evolution occurred in the human form of *FOXP2* through positive selection at the species level, since all humans carry these two mutations. These results have led a large part of the scientific community to consider that this gene may be involved in the acquisition of the articulated language so typical of our species. One fact is worth underlining: when these two human-specific mutations are inserted into the *FOXP2* gene of the mouse, we observe in these 'humanized mice' changes in dopamine levels, neuronal morphology, synaptic plasticity and, curiously . . . the squeals of their babies! Even though other genes are unquestionably necessary for the establishment of articulated language in humans, the involvement of *FOXP2* in this complex phenotype seems increasingly clear.

Local adaptation of populations

As we have already seen, nature does not make jumps. Nor does it give the biologist any gifts: its ways are many, and they are complex – a tangle that is difficult to unravel. Behind the concept of natural selection, we discover a whole set of forms and variations, which we can now explore thanks to genetics.

Whether it is a matter of selection at species level or local adaptation, positive selection can follow very different evolutionary trajectories. Each of these forms of selection leaves specific molecular signatures, and these signatures may be detected by a constantly increasing number of statistical methods. Depending on the nature of the genetic data analysed, these tests may tell us the period when the selection occurred.

According to the so-called *classic selective sweep* model, positive selection works on a newly appeared advantageous mutation which from that point on increases in frequency until it reaches fixation. The case of *FOXP2* is a good example of a classic selective sweep at the level of the entire human lineage. But there are alternatives to this classic model, such as *selection on standing variation*, which involves the selection of an allele that is already present within a population and whose frequency will increase following an environmental change. Another possibility is *polygenic selection*: this refers to the simultaneous selection of mutations on a large number of genes. In this case, unlike the classic sweep, each mutation makes just a small contribution to a better adaptation. Finally, adaptation to the environment may also occur through *balancing selection*, a form of natural selection that preserves functional diversity across long time periods.

Classic selective sweeps within human populations, though rare, remain the most studied of these selection regimes. Whole-genome studies of different populations around the world have made it possible to detect several hundred genomic regions presenting signatures of positive selection over a period between 1,500 and 45,000 years ago. These genomic regions have been identified on the basis of characteristic molecular signatures of positive selection, like the degree of population differentiation or the length of the haplotypes. Overall, these studies point to the idea that the selection pressures associated with humans' changing environment result in some advantageous mutations being specific to a given population or geographical region. In turn, these mutations influence variable phenotypes in humans, such as height, skin pigmentation, immune response, lactose tolerance, the metabolism of fatty acids, and the levels of haemoglobin circulating in the blood.

In 2008, we decided to test the hypothesis that the phenotypic variability observed in human populations may reflect differences in local genetic adaptation. With this in mind, we used a pan-genomic approach that measured the degree of differentiation of about 2.8 million mutations in several human populations. We observed that positive selection had mainly accentuated the differences between populations as regards non-synonymous mutations and those that are located in regulatory regions. This study showed that, by increasing the genetic differentiation in certain genes, natural selection made it possible for human populations to experience local adaptations. Thanks to this approach, we managed to identify more than 500 candidate genes subject to selection pressures in certain human populations. The functions of these genes are very varied: some are involved in skin pigmentation, others in the host–pathogen interaction or the metabolic syndrome (obesity, diabetes, hypertension). This approach also made it possible for us to demonstrate at the whole-genome level that, in a significant proportion, the genetic differences between geographically distant populations can be accentuated by phenomena of local adaptation to the environment.

Subsequently, dozens of other studies using various statistical methods have tested and confirmed this hypothesis. For example, in 2013, Pardis Sabeti's team at Harvard University used the complete sequences of the 1000 Genomes Project to study the extent of positive selection in different populations of the world. They identified more than 400 genomic regions that might be candidates for positive selection, including thirty-five described as 'very robust' – corresponding to mutations that modify the protein sequence.

However, a major problem with all these studies is the presence of false positives, since a fairly sizeable number of genes that are candidates for positive selection are not replicated between the studies. In fact, establishing a complete picture of the way in which local selection has truly affected phenotypic variability in humans remains a hard task. First, we need to identify the genomic regions under selection, then the phenotypes on which selection acts, and, ideally, the external conditions leading to selection. Nevertheless, the study of local selection within populations living in extreme conditions or having a unique diet gives us the possibility of gaining a better understanding of the adaptative nature of human physiological responses to environmental pressures. In the pages that follow, I will focus on the most supported examples of local adaptation in humans, on both the genetic and the phenotypic level.

Digesting milk: an iconic case of positive selection

One of the most emblematic cases of positive selection in humans is that of the lactase gene (*LCT*), which is responsible for lactose tolerance in adults. In most humans, the ability to digest lactose, which is present at high levels in milk, decreases rapidly after weaning. However, it has long been observed that there is a great variability of this phenotype in different human populations, some still digesting milk in adulthood. A strong correlation exists between the ability to digest milk and the lifestyle of populations: populations whose lifestyle is based on agriculture and livestock farming tend to present this phenotype of lactose tolerance. Examples are the populations of northern Europe and, to a lesser extent, those of central and southern Europe, and those of the Middle East, East Africa, Central Asia and southern Asia. The correlations observed between lifestyle and lactose tolerance led researchers to advance the hypothesis that lactose tolerance has conferred a selective advantage, since the consumption of fresh milk and other dairy products permitted an efficient caloric contribution, the assimilation of calcium at high latitudes, or an increase in the absorption of milk water in arid environments. Thus, cultural innovations – such as the introduction of husbandry and farming and the use of its products – create adaptive 'opportunities' that contribute towards shaping the species biologically over the course of its evolutionary history.

Over the last twenty years, this hypothesis has been broadly confirmed. The ability to digest milk in adulthood, because of the persistence of the expression of the *LCT* gene, is indeed under genetic control: a regulatory mutation of the expression of the *LCT* gene (T-13910) was initially identified in Europeans. Since then, other mutations (C-14010, G-13915, G-13907) have been associated with the digestion of milk in adulthood in other populations of Africa and the Middle East. For example, the mutation C-14010 is the most frequent within populations of farmers in Tanzania and Kenya, while the mutations G-13915 and G-13907 are instead present in Sudan and Kenya. Interestingly, these mutations show signatures of strong positive selection in the different populations studied, confirming that lactose tolerance was a phenotypic trait that was highly advantageous to their survival, and therefore a phenotype involved in these populations' adaptation to their local environment.

One detail of major interest should be underlined: the dating of these mutations under positive selection, performed by Sarah Tishkoff's team at the University of Pennsylvania, indicates that selection began about 9,000 years ago for European populations and 3,000 to 7,000 years ago for African populations. These dates are totally consistent: they would mean that selection began at a period when these populations started to adopt cultural practices linked with agriculture and animal husbandry. This is a striking example of convergent evolution, an evolutionary mechanism that leads to the acquisition of the same phenotype through independent mutational events. The same cultural practices, the same selection. This is, therefore, also an example of local adaptation resulting from a strong selective pressure linked to shared cultural features, such as the domestication of cattle and the consumption of milk in adults.

Sun on skin: climate and human pigmentation

Quite apart from this example of local adaptation to a mode of nutrition, humans have been able to adapt more generally to a wide variety of climatic conditions all over the planet, including extreme conditions. Human populations have adjusted to life in the tropical rainforest, to the icy temperatures of the Arctic, and even to the hypoxia of high-altitude regions. How did that happen? What are the evolutionary mechanisms at work? A few years ago, Anna Di Rienzo's team at the University of Chicago began to search for mutations associated with human adaptation to different environmental variables, such as temperature, humidity, levels of sunlight and rainfall. It had long been suspected that the variability of certain phenotypes in humans – like skin pigmentation, basal metabolism (the energy requirements necessary to the survival of the organism) or the shape and size of the body – might be the result of differential adaptation to climatic conditions. By analysing the variability of the genomes of sixty-one populations around the world and correlating the frequency of mutations in their genome with different climatic variables, this team did indeed find evidence of genetic adaptation to the environment. Some of these mutations were involved in skin pigmentation, hair colour and morphology, as suspected, but also in immunity, infection and cancer. These results show, at the genome-wide scale, that adaptations to climate have shaped the phenotypic variability we currently observe in humans.

The variation in skin pigmentation is one of the most striking examples of phenotypic diversity. Unlike that of other primates, human skin is not covered with a thick layer of hair and is in direct contact with the environment. Exposure to the sun's rays is the main driving force behind the evolution of pigmentation in humans, with strong selection pressures favouring a darker pigmentation in low latitudes, given the obvious protection it offers against photodamage (for example, melanomas), and above all to avoid the protection that it offers against the photodegradation of folate, a metabolite essential to the development of the embryonic neural tube and to spermatogenesis. On the other hand, selection has favoured a lighter pigmentation in higher latitudes in order to maintain the production of vitamin D, synthetized in the skin under the effect of ultraviolet rays, in regions where the levels of UV irradiation are weaker.

Over the last decade, several genes associated with the variation in skin colour have been identified (for example *OCA2*, *TYRP1*, *TYR*, *SLC24A5* and *SLC45A2*), of which some present signatures of positive selection. In addition, a recent analysis of 230 ancient samples living at different periods between about 2,000 years and 8,000 years ago made it possible to observe that a genetic variant associated with lighter skin probably reached its point of fixation in Europeans relatively recently, in the course of the last 4,000 years.

Although many mutations associated with skin pigmentation have been identified, most studies deal with European populations. Unfortunately, this is a fairly general rule in human genomic studies, and, consequently, the genetic basis of skin colour in other regions of the world remains largely unknown. Within Africa, the variability of skin pigmentation is immense, and the range of shades is very extensive: it goes from the dark skin of the Nilotic shepherds to the fairly pale skin of the San hunter-gatherers of southern Africa. Two recent studies, one by Sarah Tishkoff of the University of Pennsylvania and the other by Brenna Henn of the University of California at Davis, have tackled this question and identified new genes linked to skin pigmentation in Africa. They have also shown that the genetic architecture of skin pigmentation in Africa is much more complex than that in Europe, where a handful of mutations explain most of the variability of the phenotype. In addition, the geographical distribution of the African mutations responsible for skin pigmentation seems to be the result of genetic adaptation to the different latitudes. In some cases, these mutations may be introduced by means of migrations, as is the case with a mutation associated with a lighter skin (this concerns the *SLC24A5* gene), introduced into East Africa by gene flow from Eurasia in the last 3,000–9,000 years.

All these studies point to the idea that skin pigmentation is a phenotype that has been strongly influenced by natural selection in humans. Conversely, the role of natural selection in the evolution of hair and eye pigmentation, the genetic basis of which is also increasingly known, seems to have been negligible. Other factors probably played a major role in the current distribution of these phenotypes, such as specific demographic events or genetic drift.

Sexual selection may also play a role in the evolution of human phenotypes. It is defined as a 'sex-dependent' competition, associated with the choice of partner, which privileges certain physical or morphological characteristics, such as hair or eye pigmentation. Darwin believed firmly that sexual selection was a major force shaping the phenotypic differences between different human populations. However, the role played by sexual selection in the distribution of hair and eye colour across human populations has not yet been explored in specific detail. As for its influence on skin pigmentation, it does not seem to have played a major role but has probably accentuated the degree of sexual dimorphism (i.e. the difference in appearance of the male and female of the same species) in the skin colour of certain populations. Some studies show that men prefer lighter-skinned women in some East Asian and South Asian cultures, which probably contributed towards accentuating the generally lighter pigmentation observed in women because of an increased need for vitamin D during pregnancy and breastfeeding.

Cold and the omega-3

But there isn't only the sun. There's also the cold . . . The Arctic environment must be one of the most extreme on earth, with its low temperatures and lack of sunlight. It would be an adaptive challenge for any living creature, and humans are no exception. Despite that, several cultures living on the continents concerned have managed to adapt to this inhospitable environment, such as the Chukchi and the Evenks in Siberia, the Samis of northern Europe and the Inuit of North America and Greenland. These populations have had to adapt not only to the extreme cold, but also to a marine diet rich in polyunsaturated omega-3 fatty acids, which are found in large quantities in seal meat, whale meat and cold-water fish. Up until very recently, the genetic basis of adaptation to the cold and to this unusual diet remained unexplored. Recent studies have tackled this question, studying the genomic diversity and signatures of natural selection in populations living in the far north.

An initial study in 2014 focused on the Chukchi, Eskimo and Koryaks in eastern Siberia. By sequencing their genomes and comparing them to those of other populations in the Arctic Circle, the scientists identified a non-synonymous mutation in the *CPT1A* gene, present at high frequencies in these populations (almost 70%) and displaying strong signs of positive selection. The frequency of this mutation, which presents one of the strongest selection advantages ever identified in a human population, probably increased rapidly in the course of the last 6,000–23,000 years. The same mutation is present in the Inuit of Canada and Greenland at a frequency of about 50%, but, surprisingly, it is associated with hypoglycemia and a high rate of infantile mortality.

Why should a mutation associated with deleterious effects that have a direct impact on survival be the result of positive selection? And how could it have reached such high frequencies? The answers can be found in the function of the protein encoded by the *CPT1A* gene, which is to import long-chain fatty acids into the mitochondria to be used later in their oxidation. So this mutation, even though deleterious, probably gave a metabolic advantage to the populations of the Arctic Circle in relation to their traditional fat-rich diet. Later, we will see other examples of this evolutionary paradox or mismatch, in other words, the collateral damage of natural selection in the past on the health of individuals today.

Similarly, two independent studies explored how the Inuit of Greenland have adapted to their environment. The first was focused on the study of the genetic basis of type 2 diabetes, the frequency of which is particularly high within this population. By conducting an analysis of genetic association, it identified a mutation in a gene (*TBC1D4*) that plays a role in diabetes and high rates of circulating glucose and insulin. This mutation presents a frequency of about 20% in the Inuit but is practically absent in other populations around the world. Even though it is tempting to think that adaptation to the traditional hypoglycemic diet of the Inuit favoured the increase in the frequency of this mutation, which also affects the absorption of glucose, it remains an open question whether this is due to selection or to genetic drift.

This is the question that was tackled more directly in the second study, which studied how natural selection affected the Inuit's adaptation both to a cold climate and to a lipid-rich diet. At the whole-genome level, the genomic region presenting the strongest signatures of positive selection is located around a group of genes involved in the chain of desaturation of the fatty acids (*FADS1-FADS2-FADS3*), followed by two genes (*WARS2* and *TBX15*) involved in the waist–hip ratio and in the differentiation of the adipocytes, which suggests a possible involvement in the adaptation to cold. As for the selective pressure underlying the signal of selection detected in the *FADS* genes, it seems logical to suppose that it occurs in response to the Inuit's diet rich in fatty acids. It is even more interesting to observe that these same genes have a major effect on height and weight, as well as a protective effect on the rate of cholesterol and triglycerides. This example shows clearly how pleiotropy (i.e. the fact that the same gene may be causally involved in several phenotypes) can have major effects on the phenotypic features of a given population. Natural selection has favoured mutations in the *FADS* genes in response to the Inuit's diet, but the phenotypic consequences go well beyond the metabolizing of fatty acids and lead to other phenotypic traits, such as shorter height.

In 2018, another study examined adaptation to cold more directly, looking at the evolutionary history of the *TRPM8* gene – a gene that encodes for the only known the only known temperature receptor that is involved in the detection of, and reaction to, cold through physiological thermoregulation. This gene harbours a mutation in its regulatory region that is found in populations living in northern latitudes and presents signals of positive selection. Two elements point to the adaptive nature of this mutation. First, it presents a strong difference in frequency depending on the populations under consideration. Second, its geographical distribution depends on latitude and temperature: its frequency varies from 5% (in Nigerians) to 88% (in Finns).

Nevertheless, this is not a regime of positive selection through a classic sweep, but rather a phenomenon of selection on standing

variation. The selected mutation at the *TRPM8* gene probably originated in Africa, where it evolved in a neutral fashion. It was in Eurasia
that it reached high frequencies, with positive selection beginning
about 25,000 years ago, coinciding with the Last Glacial Maximum.
The comparison of these data with those derived from ancient DNA
indicates that the mutation was present in Europe at very high frequencies in the last 3,000 to 8,000 years. But there is a price to be paid
for such an adaptation to cold: the mutation is also associated with
greater vulnerability to migraine, which partly explains the higher
incidence of this condition in populations of European origin.

Growing up in the rainforest: humidity and height

To complete this tour of the planet's inhospitable climates, it is now
time to talk about the tropics. The word inspires dreams, the climatic
reality is sometimes far from that. If humans have adapted to the
extreme climates of the northern latitudes, they have also faced hot,
humid climates, and their genomes bear the marks – another subject
that has fascinated anthropologists and geneticists over the last decade.
The tropical rainforests are one of the most hostile environments in
the world, encompassing high temperatures, high levels of humidity
and a strong prevalence of infectious agents. The inhabitants of the
rainforest often have a shorter lifespan, which directly affects their
reproduction and consequently results in a major selection pressure.

The most distinctive phenotype of populations inhabiting the
rainforest is their short stature (less than 1.50 metres in adult males).
This phenotype is present in some populations of hunter-gatherers
across Africa, but also in Asia and South America. Among these
populations, the Efé, hunter-gatherers living in the Ituri Rainforest
in the Democratic Republic of the Congo, stand out: they are the
shortest of all human populations, with an average of 1.36 metres in
women and 1.43 metres in men.

How do we make sense of the convergent origins of shorter
stature in some groups from Africa, Asia and South America? They
may be thought of as adaptive responses to the almost identical

environmental challenges posed by the rainforest to populations who encounter them in different areas of the planet. Several hypotheses have been advanced to explain short height as such a response: limited access to food resources, protection against the overheating associated with extreme humidity, greater mobility in the rainforest and/or an evolutionary compromise between arrested growth and the beginning of reproductive age, since these populations generally have a shorter lifespan. How to distinguish between these hypotheses? Over the last few years, several studies in human genomics and population genetics have shed new light on the genetic basis of shorter stature and the nature of the selection pressures leading to it.

Different teams have examined the genetic adaptation of Central African hunter-gatherers to forest life – especially those led by Paul Verdu and Evelyne Heyer of the Muséum national d'Histoire naturelle, and Etienne Patin and myself at the Institut Pasteur. They have made it possible to identify several genes or genomic regions involved in the variation in height that also present signatures of natural selection. However, these studies remain unsatisfactory, requiring further confirmation, since the genes identified often differ from one study to the next, perhaps reflecting a high rate of false positives. However, the adaptive nature of two genes in particular is increasingly appreciated: *FLNB* and *EPHB1*. Not only do these genes present robust signatures of positive selection, but their variation also affects body height in model organisms or directly in humans. It should be noted that, to explain the lack of replication between the studies, we may envisage a more encouraging hypothesis (more encouraging than uncertainty, anyway): the existence of convergent phenomena of adaptation, perhaps, like those described for lactose tolerance. The genomic data obtained by George Perry's team at Penn State University point in this direction: they show a difference in genetic adaptation to the Central African rainforest between the hunter-gatherers of the west and those of the east.

My team recently examined this subject in an attempt to gain a better understanding of the adaptive nature of the shorter stature in African rainforest hunter-gatherers and the evolutionary mechanisms involved in such an adaptation to forest life. With this aim in mind, we collected and analysed genomic data from six groups of hunter-gatherers from Central Africa, ensuring an 'interpopulation replication' of the results. We identified a single event of strong positive selection, through a selective sweep, shared by all the groups of hunter-gatherers, and which is obviously absent in their neighbours, the farmers. This event is located in a genomic region that includes the *TRPS1* gene, which is involved in certain morphological traits: delayed growth, cranio-facial morphology and hypertrichosis (i.e. excessive hair growth), but also immune response.

We also detected signals of polygenic selection associated with height and certain life-history traits, such as the onset of menstruation (or menarche) or the age of giving birth to the first child. In-depth analyses have shown that these latter traits resulted from the pleiotropic effects of the genes associated with height. In other words, the true adaptive phenotype is height, but, since the genes regulating this also have other functions, the selection of the 'height' phenotype has simultaneously altered other phenotypes, such as the age of reproduction.

It should be said that polygenic selection has also affected other biological functions in these groups of hunter-gatherers, like certain immune functions associated with the responses of the mastocytes to allergens and microbes, the interleukin-2 signalling pathway and host–virus interactions. This observation also points to a genetic adaptation of the rainforest hunter-gatherers to the high presence of pathogens in the rainforest. Overall, these discoveries support the hypothesis that the short height of the rainforest hunter-gatherers has a genetic basis, and not an environmental one (linked, for example, to food resources), and that short height was an adaptive trait for life in the rainforest.

Height in Europe: polygenic selection?

Let's leave for a moment the rainforest and find ourselves instead in the temperate climates of Europe. Here, we encounter another question of height: why are northern Europeans generally taller than southern Europeans? Is this also the result of natural selection? In Europe, hundreds of mutations in a large number of genes have been associated with variations in height, with each making a slight contribution to the quantitative trait in question. Even if these mutations, taken one by one, affect height by only a few millimetres, a slight difference in the average frequency, if it affects hundreds of mutations, may generate a difference of several centimetres in the study population.

Many recent studies have examined this phenomenon, searching for differences in the frequency of mutations associated with height compared with those that might be expected through genetic drift. They've found that polygenic selection could partly explain the differences in height observed between European populations.

In the case of the British population, based on data from the UK Biobank, it has been proposed that polygenic selection favouring an increased frequency in the mutations associated with greater height took place in the course of the last 2,000 years. However, more recent studies reach much more nuanced conclusions. The strong signal of polygenic selection detected for height could be the result of a simple methodological bias in the previous studies, the data not having been corrected by the 'stratification' of the populations – in other words, the presence of a systematic difference in allelic frequencies between several subpopulations of the same population, a difference that is independent of the trait studied. Once the studies are corrected by this key factor, the signal of polygenic selection linked to height appears much more moderate.

Mountain sickness: hypoxia at altitude

Let us now leave Europe and return to observe an example of adaptation in the face of a truly hostile environment: life at high altitudes. At an altitude of 4,000 metres, the amount of oxygen present in the air is reduced by 40% compared with at sea level. Yet more than 25 million people live at altitudes above 3,000 metres, where they are

permanently subject to conditions of hypoxia, that is, to an insufficient provision of oxygen compared with the needs of the organism.
Of course, at lower altitudes, the body can acclimatize to an increasing scarcity of oxygen to a certain extent, thanks to physiological
mechanisms. Higher up, things are different. There are at least four
populations living at high altitudes for whom there exists evidence
of genetic adaptation to hypoxia: the Tibetans of the Himalayas,
the Quechuas and the Aymaras of the Andes, and the Amharas of
the Simien Mountains in Ethiopia.

An initial study of the Tibetans compared their genomes to those
of populations living at low altitudes like the Han Chinese. The
researchers identified a gene known as *EPAS1* which presents several mutations at a frequency of 85% in the Tibetans and only 5% in
the Han, despite the rest of the two populations' genomes being
very similar. This gene encodes a protein that is part of the family
of *hypoxia-inducible factors* (HIFs), which regulate the cellular
response to hypoxia. HIFs generate an increased secretion of
erythropoietin, which favours the production of red blood cells,
thus increasing the concentration of haemoglobin and therefore of
oxygen in the blood. A point worth noting is that the mutation
involved in the genetic adaptation of Tibetans to hypoxia is also
found in the genome of the Denisovans. This suggests that the
beneficial variant first appeared in the Denisovans and was later
transmitted to modern humans through admixture. A later chapter
will deal in detail with the adaptive nature, in modern humans, of
this so-called 'archaic' admixture.

Other genes that are candidates for being involved in the response
to hypoxia have been identified in populations living in the Andes
and the Ethiopian highlands. We observe in them potentially beneficial mutations in genes belonging to the family of the HIFs (like
EGLN1 or *BHLHE4*). We also find in these populations variants in
genes that do not belong to the HIFs. As in the case of the digestion
of milk in adulthood, the adaptation to hypoxia is a remarkable
example of convergent evolution, where different mutations, different genes and different mechanisms lead to the same phenotype

that allows for adaptation to the lack of oxygen characteristic of high-altitude regions.

Hypoxia in apnea

It is not necessary to live high in the mountains to lack oxygen. Apnea, the temporary cessation of breathing, also leads to hypoxia. On average, one can hold one's breath for several dozen seconds, occasionally more than a minute, rarely beyond that, without training. But there is a particular population whose members are able to remain in apnea . . . for thirteen minutes! It is the Bajau, a people living in the Philippines, Indonesia and Malaysia. They are known as 'sea nomads'. In order to catch the fish and crustacea on which their lives depend, the Bajau spend about 60% of their day diving to a depth of about 60 metres, which corresponds to conditions of extreme physiological stress. This unique adaptation allowing them to take prolonged and repeated dives has for a long time been considered by the scientific community as being an example of simple 'phenotypic plasticity', in other words, the ability, independent of the genetic code, to express different phenotypes depending on the environmental conditions.

A study conducted by Melissa Ilardo and Rasmus Nielsen of the University of California in Berkeley determined that it did in fact have a genetic basis. Genomic analysis of the Bajau of Indonesia revealed that they have adapted to acute hypoxia through a hypertrophy of the spleen. This organ contracts in response to apnea and provides the necessary supply of oxygen by the expulsion of red blood cells. The spleen of the Bajau is larger than that of the Saluan, a population living nearby who do not practise diving. The difference in volume is observed in all Bajau, including those who do not dive, which contradicts the idea of phenotypic plasticity. In addition, this specificity appears to be due to a variant in the *PDE10A* gene, which is involved both in the thyroid function and in the size of the spleen. Since the thyroid hormones regulate erythropoiesis in the course of postnatal development, a larger spleen in the Bajau may be the result of a higher volume of erythrocyte cells.

The fact that the mutation in question presents signatures of positive selection in the Bajau suggests a genetic adaptation in response to their lifestyle, favouring an increase in the quantity of oxygenated cells. This study has thus revealed a new physiological mechanism of adaptation to extreme environmental conditions in humans. The case of the Bajau also provides a clear example of genetic adaptation to cultural practices, as in the case of the consumption of milk associated with the practice of livestock farming. These examples illustrate the co-evolution of culture and biology in the course of the last few thousand years of our history. We will return to this topic in detail.

And adaptation continues . . .

We have seen a few examples of phenotypic traits that seem in the past to have provided selective advantages for human survival in specific environments. To what extent have these selection pressures persisted into more recent times? This is the specific question that Jonathan Pritchard's team at Stanford University attempted to answer in 2016. To detect positive selection, most available statistical methods have been conceived to measure it over long periods of time. Pritchard's team, conversely, developed a new method, called the *singleton density score* or SDS, which makes it possible to detect rapid and recent changes in the frequency of mutations, and so to measure selection over the last 2,000 years. This method, though, requires data from whole-genome sequences, which had been rather costly but have recently become more affordable thanks to developments in genomic technologies.

By applying this method to more than 3,000 whole genomes of British origin, they identified several variants that seem to have continued to be beneficial during the last two millennia. Some notable examples are the allele associated with lactase persistence, which indicates that the benefit of this variant has persisted until recently. The SDS method has also detected strong signals of positive selection in the genomic region of the human leukocyte antigens (or HLA, which corresponds, as we have seen, to the major histocompatibility complex in humans involved in adaptive immunity). This is also the case with genes involved in pigmentation, favouring blonde hair and blue eyes, probably through sexual selection.

This study also identified some cases of polygenic selection that have been 'active' until recent times. For example, some genes favour an increase in height, circumference of the head of the infant and weight at birth, or else sexual maturity in women, while others favour a decrease in the rates of glycated haemoglobin or in the body mass index, exclusively in men. However, as we have seen with height, given the potential biases associated with population stratification we must consider these results with caution. Apart from this methodological bias, the interest of this study lies in the ability to identify events of genetic adaptation shaping the phenotypic diversity of human populations in very recent periods.

Balancing selection: the benefits of diversity

There is finally a type of natural selection we have not yet spoken about: balancing selection, which maintains diversity within a population or a species thanks to several mechanisms. The *heterozygote advantage* is part of it: this refers to the advantage of heterozygous individuals, that is, those who possess two different forms of the same gene, compared with homozygous individuals, who present two identical copies of the same gene. Balancing selection may also manifest itself in the form of *frequency-dependent selection*, in other words, the selective value of a phenotype depends on its frequency compared with other phenotypes in the population. For example, an individual having a relatively rare phenotype compared with other individuals will have greater fitness thanks to this rarity: this is called *negative* frequency-dependent selection.

Unlike other forms of selection, balancing selection may maintain functional diversity over long periods, even millions of years if the conditions of selection remain constant through time and are strong enough to avoid the loss of mutations due to genetic drift. In some cases, these 'balancing' mutations may even survive events of speciation, leading to what is called a 'trans-species' polymorphism.

Humans and chimpanzees against Brussels sprouts

The perception of bitter tastes has long been considered a case study of ancient balancing selection. Sensitivity to bitterness provides animals, including humans, with an important tool for interacting with their environment, making it possible for them to detect various toxic compounds in foods. In 1931, a chemist named Arthur Fox noted that some people found phenylthiocarbamide (PTC), an organic compound present in many plants, including broccoli, Brussels sprouts and pepper, very bitter, whereas the rest of the population was insensitive to the taste. The genetic basis of this detection skill was identified seventy-three years later: three non-synonymous variants in the *TAS2R38* gene differentiate the 'tasting' from the 'non-tasting' phenotype.

The ability to perceive the taste of PTC is a dominant genetic trait that constitutes a classic example in human genetics. The distribution of tasting and non-tasting phenotypes is observed at intermediate frequencies in most populations, which is consistent overall with a scenario of balancing selection through heterozygote advantage, or the advantage of diversity. Nevertheless, we observe a certain variability across populations: the tasting phenotype varies from about 75% in South-East Asia and the Pacific to about 50% in Europe, to almost 100% in Native Americans.

In 1939, Ronald Fisher reported that chimpanzees, like humans, presented both tasting and non-tasting phenotypes, at relatively equal frequencies. He interpreted this observation as an ancient polymorphism that appeared before the human / chimpanzee divergence and has been maintained by balancing selection during the last 5–6 million years. The publication of the chimpanzee genome in 2005 opened up the possibility of exploring their genetic diversity. Stephen Wooding and Michael Bamshad, at the University of Utah at the time, revisited this classic example of 'shared' balancing selection in 2006. Through the sequencing of the *TAS2R38* gene in eighty-six chimpanzees and comparing it with the human version, they made an unexpected discovery: Fisher's interpretation was wrong. Unlike in humans, the non-tasting phenotype in chimpanzees is distinguished from the tasting phenotype by a single variant, which does not correspond to any of those observed in humans. That might mean that, contrary to Fisher's prediction, the perception of bitterness evolved independently on at least two occasions in the course of the evolution of great apes. This would make it a case of convergent evolution rather than an ancient event of balancing selection shared by the two species.

Is balancing selection a frequent mechanism?

In other cases, balancing selection takes place within species because of specific environmental pressures. A good example of recent balancing selection, understood since the 1950s, is the mutation known as HbS in the gene encoding haemoglobin. It is one of the most striking examples of natural selection maintaining a high frequency of a deleterious mutation in the human population, because of the protection it confers (about ten times larger, in the heterozygous state) against malaria caused by the parasite *Plasmodium falciparum*. On the other hand, the homozygous carriers of two copies of HbS develop sickle-cell disease, an often-fatal form of anaemia caused by deformations of the red blood cells.

According to recent whole-genome studies, balancing selection, long considered a rare event in humans, may be more widespread than was once suspected and seems to have affected a fair number of genes, especially genes associated with immune functions and host–pathogen interactions. Just like access to food or life in extreme conditions of cold or hypoxia, pathogens have been a major driver of adaptation in humans, perhaps even the most important during our evolution. The imprint of the selection pressure exerted by infectious agents on our genomes is so important that we have deliberately avoided talking about it here in order to devote the whole of the following chapter to this topic, which is a crucial concern in evolutionary biology, human immunology and medical genomics.

4.

Humans and Microbes

Genetic adaptation is not only a response to environmental challenges such as climate and altitude. Like all living creatures, we have had our predators and they have contributed towards making us what we are. Similarly, like all living creatures, we have been confronted by many pathogens, many of which cause life-threatening infectious disease, and we are regularly confronted with new ones. We have had to deal with them, and our genomes bear the traces. In this part, I will focus mainly on microbes and our dangerous liaisons with them. We have learnt that the immune defences we have at our disposal today are inseparable from our evolutionary past: it is there that the solutions we have inherited can be found, like the genes of those who, in the course of this history, have survived the great epidemics of the past, such as plague, tuberculosis and cholera. Here, too, the tools of genomics bring new perspectives.

Alice and the Red Queen: an arms race

'Now, here, you see, it takes all the running you can do, to keep in the same place. If you want to get somewhere else, you must run at least twice as fast as that!' This is the answer the Red Queen gives Alice in a passage from Lewis Carroll's *Through the Looking-Glass*, where they both throw themselves into a mad race. This metaphor has been used to symbolize the 'arms race' between the species, that is, their need to constantly evolve faced with competition and environmental changes. This competition leads the competing species into a permanent one-upmanship of adaptations and counter-adaptations. It is this permanent competition between humans and pathogens that is described in the Red Queen hypothesis, put forward in 1973 by Leigh Van Valen in 'A New Evolutionary Law' in the journal *Evolutionary Theory*.

Humans need to be permanently evolving in order to keep pace with the species with which they co-evolve, particularly infectious agents. Those pathogens that have an impact on the reproductive success of the host because they threaten his/her health or bring about premature death may lead to the selection of mutations that increase the host's resistance to infection. Selection should therefore be more visible for pathogens that have had a long relationship with humans, like those causing perennial diseases such as malaria and cholera, but also smallpox, tuberculosis and leprosy.

Humans and microbes maintain a permanent, double-edged relationship. They complement each other, as gut microbiota can attest, but some micro-organisms are pathogenic and cause infectious diseases. Just as famines and wars have constantly burdened humans with a high rate of mortality, infectious agents has always been a major threat to our survival. Pathogens have been with humans since their appearance in Africa about 200,000–300,000 years ago, during their subsequent dispersals across every continent over 60,000 years, and during more recent cultural transitions, like the emergence of agriculture and sedentism about 10,000 years ago, or during European colonization and contemporary globalization. These migrations and cultural changes have exposed humanity to new pathogens: the increase in population density and closer contact with disease-carrying animals, including those that have been domesticated such as dogs, pigs and poultry, or animals that cohabit with sedentary humans, such as rodents.

Despite medical progress, infectious diseases continue to inflict a heavy death toll. This is the case for more recent threats such as the Spanish flu early in the twentieth century or emerging contemporary diseases like AIDS, known for less than 100 years, the Zika virus, which led to an epidemic in 2016; and different forms of severe acute respiratory syndrome (SARS), associated with coronavirus infections, which appeared in the twenty-first century and have been much spoken about during the COVID-19 crisis that began in 2019. Some pathogens cause acute diseases, like cholera, but once cured, they constitute a minimal threat. Others, on the contrary, may cause chronic infections that have a big impact on the metabolic state of individuals, such as their growth or their fertility: this is the case with the pathogens causing malaria, tuberculosis and certain parasitic infections. There are two solutions: to leverage the genetic means of evolution that can maintain our ability to defend ourselves against changing threats – but at a high cost in both time and human lives – or to rely on the means developed by modern science and medicine. These constitute an unprecedented weapon – remarkably effective, even though still imperfect – in the competition with pathogens.

Infectious diseases: mortality and natural selection

The rates of mortality due to infection, and therefore the selective pressure exerted by pathogens, remained extremely high until the end of the nineteenth century and the beginning of the twentieth, a period in which hygiene conditions improved and vaccines and antibiotics began to make their appearance. It is worth recalling that at the end of the nineteenth century only 35% of Europeans reached the age of forty. This figure suffices to give an idea of the heavy infectious burden on our species throughout its history.

Louis Pasteur and germ theory

Louis Pasteur, one of the fathers of the germ theory of disease, lost three of his daughters to what at the time was called 'fever'. With hindsight, it's clear that they probably died of an infectious disease. This was representative of most families at the time: it was not unusual that at least half of a given group of siblings would die of an infection. The question that intrigues the geneticist about that selective mortality is: 'Why that half?' In Pasteur's family, his son and one of his daughters survived into adulthood, despite likely exposure to at least one of the microbes that killed their sisters. It is therefore possible that the three girls who died carried a form of genetic susceptibility predisposing them to infectious diseases.

The advent of the germ theory of disease, which asserts that many diseases are caused by micro-organisms, led to dazzling progress in medicine and hygiene and a considerable improvement in living conditions. It was towards the middle of the nineteenth century that Louis Pasteur, the pioneer of microbiology, provided empirical evidence supporting this theory, of which there had been precursors for a long time, mostly at a level of intuition or hypotheses. The question was very controversial in Pasteur's time. Yet he demonstrated definitively that the fermentation and growth of micro-organisms were not due to spontaneous generation, a major alternative theory at the time. He also discovered that pébrine, or pepper disease, was caused by a microscopic organism that today bears the name *Nosema bombycis*. It was with the work of the German physician Robert Koch, in the 1880s and 90s, that germ theory reached its apex: it was discovered then that anthrax and tuberculosis, among other diseases, were caused by pathogens. Koch is particularly famous for having discovered Koch's bacillus, or *Mycobacterium tuberculosis*, the causative agent of tuberculosis.

Curiously, Charles Darwin makes no mention in his work of infectious diseases as being a major selective force in human evolution, even though he was a contemporary of Pasteur and Koch and so their discoveries. It would not be until the 1950s that, as we have seen, J. B. S. Haldane, Giuseppe Montalenti and Anthony Allison established for the first time a causal link between infectious diseases and natural selection. In their work, they put forward the hypothesis that diseases due to anomalies in the red blood cells, like thalassemias and sickle-cell anaemia, could protect against malaria.

The human genetics of infectious diseases

Since then, a flood of studies in human genetics have broadly demonstrated that the genetic variability of the human host may explain, at least in part, the differences we observe between individuals faced with infection. We are referring here both to an individual's susceptibility to the infection itself and to their susceptibility to developing the disease and its degree of severity. In the course of the last decade, with the advent of new genomic technologies, genetic studies have provided many examples of genes responsible for differences in susceptibility to infectious diseases, both rare and common. For example, the studies conducted by the team led by Jean-Laurent Casanova and Laurent Abel, of Rockefeller University in New York and the Institut Imagine in Paris, have revealed the power of exome sequencing to dissect the immunological mechanisms underlying the pathogenesis of some rare but life-threatening infectious diseases known as Mendelian diseases, such as herpes simplex encephalitis and Mendelian susceptibility to mycobacterial disease. Similarly, genome-wide association studies (GWAS) have been carried out on viral, bacterial and parasitic diseases. These studies have made it possible to identify genes in which variation is associated with human susceptibility to certain common diseases, such as hepatitis C, leprosy and malaria.

Given that pathogens have inflicted a heavy death toll on us throughout our history, it should be no surprise to learn that studies in population genetics have convincingly shown that human genes involved in the immune response or in host–pathogen interactions are preferential targets of natural selection. These studies have increased our understanding of the way in which infectious agents have shaped the diversity of our genomes for thousands of years. Thus, approaches based on population genetics and evolutionary biology have become an indispensable complement to more classical approaches in clinical and epidemiological genetics. They allow us to identify and understand the factors that are behind our current differences when confronting microbes, and to evaluate their relative contribution. As we will see, genetics has yet to reveal all of its many resources and place them at the service of knowledge and medicine.

Human and chimpanzee responses to pathogens

Despite the significant similarity in the genomes of humans and other primates, like the chimpanzee, genomic studies show that the different species of primates can adapt differently to the pressures imposed by pathogens. Even within our own species, different human populations have also adapted in various ways to the local pathogens they have encountered as they dispersed across the globe. It is now possible to identify the ways in which natural selection has affected the immune system – whether between different species or within our own. These go from high gene conservation preserving the status quo to strategies leveraging the advantages of diversity. We are accumulating information on the evolutionary and immunological relevance in host defence of the different human genes as well as on their role in susceptibility to, or the severity of, immune-related diseases.

The response of humans to the attacks of pathogens is not always of the same nature as the one adopted by non-human primates. There are in fact many differences between humans and other primates, in both the prevalence and the severity of diseases. Humans are more seriously affected than chimpanzees, for example, by HIV, as regards both infection by the virus and its evolution towards AIDS. Similarly, in the case of flu, *Plasmodium falciparum* malaria, and late complications associated with hepatitis B and C, humans are more seriously affected than chimpanzees. Comparative studies between species show that some of these differences can be traced back to evolutionary history, which might differ between primates depending on the species concerned. This knowledge of the genomic history of the species may, in turn, prove valuable in helping us to identify human genes that play an important role in our survival against pathogens.

One way of exploring these differences between species is to look for genes that show an evolutionary acceleration in a given lineage, as we have seen with the *FOXP2* gene associated with language in the human lineage. In 2005, two studies were conducted by Rasmus Nielsen and Carlos Bustamante, at the time at Cornell University in Ithaca, New York State, under the direction of Andrew Clark, one of the fathers of modern population genetics. They set in motion a strategy to look for such an acceleration. By comparing the degree of diversity of more than 10,000 genes between humans and chimpanzees, they found that genes associated with the immune response have been recurrent targets of positive selection ever since humans and chimpanzees started to diverge. Functions related to host defence against pathogens appear at the top of the list, followed by other functions, like smell and spermatogenesis. In addition, it appears that several genes presenting strong signatures of positive selection are involved in the immune response to viruses, for example the genes encoding interferons – a family of proteins that set off reactions allowing resistance to viruses. Given the speed at which viruses evolve, an arms race between these pathogens and the host cells may explain, in the host, a strong selection favouring new mutations in these genes.

In 2010, Luis Barreiro – at the time a graduate student in my laboratory, now a professor at the University of Chicago – and I gathered the data from several comparative studies, in order to accurately identify immune-related genes whose evolution presented an acceleration in either humans or chimpanzees or both. We detected eighty-four immune-related genes falling into these categories: seventeen genes positively selected exclusively in the human lineage, fifty-nine only in the chimpanzee lineage and eight in both. Their rapid evolution makes them excellent candidates to explain the different ways of responding to infection in these two species. One fact is worth underlining: among the eighty-four genes presenting a rapid evolution, thirty encode proteins that interact with HIV. Surprising? Given the recent nature of this infection, these selective events are probably the result of past selection pressures exerted by other pathogens; for example, ancient retroviral infections that may have set off immune response mechanisms similar to current mechanisms of response to HIV. Functional changes in these genes may also explain, at least partly, why chimpanzees, unlike humans, do not present progression towards AIDS-like syndromes after infection by HIV or by the simian immunodeficiency virus.

As with HIV, there are important differences between humans and chimpanzees in their responses to malaria. Unlike humans, chimpanzees are not subject to infections from *Plasmodium falciparum*, and, even though they are infected by *Plasmodium reichenowi*, the disease does not develop into severe cases. The rapid evolution we observe in the glycophorins A and C – proteins involved in the invasion of the red blood cells by *Plasmodium falciparum* – may explain the greater resistance to malaria observed in chimpanzees.

But if we want to investigate how genetics contributes to the differences observed between primate species when faced with infection, we must look beyond the differences in their genomic sequences, which, since their divergence, is only slightly above 1%. Another avenue to follow would be to focus on their gene content, in other words the gains or losses of genes specific to each species. Among gene families that show a rapid renewal in terms of gains or losses, those that are involved in the immune response are once again over-represented. One pertinent example is the family of immunoglobulins and the HLA genes – pillars of immunity – where we observe important differences, in terms of gene content, between humans, chimpanzees and macaques.

Overall, these observations suggest that natural selection has played an important role in the differentiation of certain genes, or certain families of genes, linked with immunity, depending on the species of primates, which may explain the differences we observe between these species when it comes to the immune response and the clinical manifestation of certain infectious diseases.

Footprints of past pathogens on our current genomes

Over the past few years, a plethora of studies have explored the way in which natural selection has shaped the genomes of the human species. One insight seemed certain: the most obvious cause of selection underlying the levels of diversity observed in immunity genes is the presence of pathogens. This hypothesis has been formally tested by several studies that have looked for correlations between the genetic variability of human populations and the richness in pathogens of different geographical regions.

An initial study, conducted by Franck Prugnolle of the IRD in Montpellier, examined the HLA complex, which presents a remarkable level of genetic diversity, attributed to the effects of balancing selection. The study revealed that, in populations living in regions presenting a great diversity of pathogens, we also observe a larger variability in certain class I genes in the *HLA* region, especially at the level of the *HLA-B* gene. Subsequently, many studies adopting the same approach have shown that other genes associated with host defence, like the antigens of the blood groups or the interleukins, present levels of diversity correlated with the presence of specific pathogens, like viruses, protozoa and parasitical worms. Of course, there is an implicit hypothesis in these studies that the richness in pathogens has remained unchanged over time, and they do not take into account their degree of virulence or pathogenicity. Nevertheless, their results support the idea that the selection pressure underlying the diversity observed today in the immune-related genes is in fact exerted by the presence of pathogens.

Of course, some of the correlations observed between human genetic diversity and pathogen presence could be due to other environmental variables, themselves associated with the presence of pathogens, such as climate, diet and lifestyle. It was this question that Matteo Fumagalli and Rasmus Nielsen of the University of California in Berkeley investigated in 2011. They studied the correlations of the allelic frequency of nearly 500,000 genetic variants in fifty-five human populations with a set of ecological variables describing each geographical location: climatic variables (temperature, humidity, rainfall, amount of sunlight, etc.), lifestyle and diet (hunting, gathering, agriculture, fishing, livestock farming), and pathogen load (viruses, bacteria, protozoa and helminths, or parasitic worms).

Among all these variables, the scientists observed that pathogens remain the principal driving force in the local adaptation of populations. In fact, they identified 103 human genes whose genetic diversity is strongly correlated with that of pathogens, after correction for the effects of the demography of the populations. They also showed that these genes are enriched in biological functions that are relevant in this context, like the inflammatory response and the pathway of the *Toll-like receptors* (or TLRs), a major pathway of innate immunity involved in the early recognition of pathogens and the establishment of an efficient response to eliminate them. In addition, it seems that parasitic worms, or helminths, have imposed stronger selection pressures on the host than other infectious agents. This may be due to the slower evolution of helminths, resulting in a more stable geographical distribution.

Another study, using a radically different approach based on the analysis of inter-species data, has suggested that viruses are the infectious agents that have had the greatest impact on the evolution of mammals, shaping the diversity of a large number of human proteins. Whatever the specific pathogens that lead to the selection pressures observed in humans, these studies add weight to the idea that the diversity of human genes involved in the immune response has been shaped by human exposure to infectious agents in the course of evolution. Our enemies have shaped us; that which threatens us has also transformed us.

A preciously preserved heritage

It is still difficult to identify the specific pathogens that are behind the patterns of selection observed and to determine their causal role. Despite these limitations, studying the way natural selection has affected our genomes, as well as the extent and form of its action, has proved crucial in identifying the immune-related genes and the immune functions in regarded to which the maintenance of diversity, or the favouring of novelty, has been beneficial to the host. In the context of balancing selection, we have seen that the levels of genetic diversity may be maintained over long periods of time, even surviving speciation events.

We find an example in the remarkable diversity of the major histocompatibility complex in vertebrates, or HLA in humans. In this case, the genetic diversity has been inherited from distant ancestors, such as other primates, but also from other mammals, as well as birds and fish. Similarly, it has been shown that the ABO system of blood groups is a trans-species polymorphism, shared by different species of primates, including humans. Putting aside pressures linked to pathogens, it seems that other factors, like sexual selection, have contributed to the maintenance of the trans-species diversity observed in the HLA and ABO systems.

In fact, we have known for a long time that the HLA system, which helps the immune system to distinguish the 'self' from the 'non-self', also plays an important role in the choice of a mate in many species of vertebrates. The explanation for this phenomenon relies on the hypothesis that by selecting a mate having HLA genes different from one's own we ensure that the offspring will inherit a varied immune repertoire, which should increase their chance to resist infection. In the case of humans, the studies of Raphaëlle Chaix (Muséum national d'Histoire naturelle of Paris) and Carole Ober (University of Chicago) have shown that the choice of a mate is indeed influenced by the variability of the HLA genes: it seems that the choice privileges (unconsciously) a partner with different HLA genes. What emerges from this study based on Europeans and Hutterites – a Christian Anabaptist ethno-religious group living mainly in Canada and the United States – is an example of sexual selection in humans, affecting the choice of partner.

In 2019, another study by Raphaëlle Chaix, using larger sample sizes and populations of different geographical origins, clearly supported the hypothesis that the HLA diversity influences the choice of a mate, but in a way that is strongly dependent on the social context. Thus, even though the general rule in humans is the avoidance of partners who are genetically similar, in some populations, specific social rules, like homogamy (i.e. choosing partners from the same social, ethnic or religious group, for example), take precedence over the biological factors that intervene in the choice of mate based on their HLA status.

Putting aside the HLA system, which is a classic case of balancing selection in humans, recent whole-genome studies have shown that the immune functions in particular have been affected by this selective regime. This has been tested over long periods, both in several species of primates and within our species. For example, an analysis of the genomes of humans and chimpanzees detected 125 genomic regions containing trans-species polymorphisms presenting signs of ancient balancing selection. These regions mainly comprise genes involved in immune functions (for example, the *IGFBP7* gene and genes encoding membrane glycoproteins), which suggests the long-term maintenance of functional diversity in these proteins because of the pressures exerted by pathogens in the two primate species.

Other studies, of which there are fewer, have focused on balancing selection within our species. Apart from the iconic case, presented in the previous chapter, of the HbS mutation in regions where malaria is endemic, other genes subject to this selective regime have been identified, such as the *KIR* (killer cell immunoglobulin-like receptors) genes that co-evolve with the HLA system, as well as genes encoding various proteins involved in cellular migration, host defence, or innate immunity. Although balancing selection remains a rare selective regime, it seems that it has been particularly pervasive in the case of immune responses and host–pathogen interactions, where the preservation of diversity has been particularly beneficial for survival.

Love in the time of malaria

Unlike balancing selection, positive selection acts on new mutations, or existing but previously neutral mutations, leading to an increase in their frequency in populations living in specific environments. The list of immune-related genes subject to this selection regime is constantly increasing, some of them being supported by functional or epidemiological data.

Among the most supported examples of local adaptation associated with specific pathogens or the immune response are mutations in genes that intervene in these different situations:

- resistance to malaria in Africa and Asia
- resistance to infection due to *Trypanosoma*, the cause of sleeping sickness, in Africa

- attenuated inflammatory responses in Europe and Africa
- antiviral responses involving type III interferons in Europe and Asia.

Even though these cases of positive selection are supported by evolutionary evidence, it remains difficult to identify the specific pathogen or pathogens responsible for many such events. So far, the most convincing evidence comes from pathogens that have had long-term effects on human populations. The most representative is provided by *Plasmodium falciparum*, the agent responsible for malaria: there is strong clinical, epidemiological and evolutionary evidence showing its selective effects on certain human genes. This is the case with the gene for β-globin, for which three mutations leading to changes in amino acids (HbS, HbC and HbE) confer different levels of protection against malaria. In the same way as in the case of the HbS mutation, previously mentioned, the relatively high frequencies of the HbC and HbE mutations, in West Africa and South-East Asia respectively, attest to the effects of selection exerted by malaria in these geographical regions.

Other genes have been identified that are both associated with protection against malaria and present signs of positive selection in different populations (particularly *G6PD*, *CD40LG*, *CD36* and *DARC*). The signals of positive selection may sometimes be very recent and geographically limited. This is the case with the *CD36* gene, which intervenes in the recognition of red blood cells affected by *Plasmodium falciparum*. This gene presents signals of positive selection targeting a nonsense mutation (i.e. a mutation resulting in a non-functional protein) common only in West Africans, and which appeared only in the last 3,600 years.

Another gene that is worth mentioning is *G6PD*, mutations in which can be responsible for *G6PD* enzyme deficiency. This pathology, known as favism, is the most common enzyme deficiency of genetic origin in the world. It is at the basis of what is known as 'haemolytic' anaemia, resulting from the destruction of the red blood cells. This deficiency is associated with different mutations in the

Mediterranean basin, the Middle East, Africa, India and South-East Asia – a geographical spread similar to that of malaria. Although a relationship between *G6PD* deficiency and malaria has been shown, the clinical and evolutionary links remain little understood.

In 2009, a collaboration between Anavaj Sakuntabhai's team at the Institut Pasteur and mine led us to study this enzyme deficiency in South-East Asia, due to a mutation known as Mahidol ('the surface of the Earth' in Thai), in relation to *Plasmodium falciparum* and *Plasmodium vivax* malaria. Our results showed that the Mahidol mutation reduces the parasite density of *P. vivax*, but not that of *P. falciparum*. In addition, this mutation seems to have increased in frequency in Thailand through positive selection over the last 1,500 years. This period corresponds to that in which the population that presents the highest frequency of Mahidol, the Karen, left the Himalayas for Thailand, where rice cultivation was starting to develop. The Karen would thus have started to be exposed to *Plasmodium vivax* on arriving in the region, where deforestation and agriculture favoured the proliferation of *Anopheles*, the mosquito vector.

Between cholera and plague

Apart from malaria, there are two other examples of selection probably exerted by specific pathogens at specific periods of human history. The first relates to cholera, which is caused by the *Vibrio cholerae* bacterium. It has been observed in Bangladesh, where cholera has long been endemic. Pardis Sabeti's team at Harvard University identified genes associated with the risk of developing cholera. Interestingly, these genes present at the same time signals of positive selection, which suggests that the causal agent of cholera in fact lies behind the events of natural selection observed, at least in Bangladesh.

The second study, conducted by Mihai Netea at Radboud University in Nijmegen in the Netherlands, examined two populations of different origins but living in the same environment: Europeans and the Roma people. It identified a signature of positive selection targeting a group of innate immunity genes (*TLR10-TLR1-TLR6*) that encode receptors present at the surface of the host cells and recognize microbial motifs. The fact that this selection signature is shared by the two populations has been interpreted as resulting from a convergent evolution possibly due to *Yersinia pestis*, the causal agent of the plague. However, although the plague was one of the most devastating epidemics in Europe in the course of the last millennium, and this after the arrival of the Roma people in Europe from the north of India, simple genetic admixture between the two groups would seem the simplest explanation.

These studies illustrate the impact that specific pathogens may have had on human evolution. All the same, it is hard to exclude the contribution of other pathogens exerting pressure on the same genes or the same immune pathways. This possibility is particularly critical for genes encoding receptors of innate immunity, which detect groups of pathogens sharing specific components. An additional complication comes from the interaction between human genes and non-pathogenic microbes, like those that are present in the microbiota. The latter have probably imposed their own selective pressure in order to maintain homeostasis, that is, the ability to maintain a balance of the parameters of the internal environment of the organism, like body temperature, glycemia, blood pressure and cardiac rhythm.

Mortui vivos docent: *the dead teach the living*

'The dead teach the living,' says the Latin proverb. As we have seen in the case of those mutations associated with the digestion of milk or the pigmentation of the skin, the study of DNA from ancient individuals – the dead – may teach us about the evolution of the immune system over time. The study by David Reich's team of 230 ancient individuals who lived 2,300–8,500 years ago, presented in the previous chapter, identified, once again, the same *cluster* of genes

(*TLR10-TLR1-TLR6*) as evolving under strong positive selection pressure in Europe. The hypothesis advanced to explain this fact – which needs to be further explored – is that some variants selected from among these genes conferred protection from infectious diseases, such as tuberculosis and leprosy. The same study detected strong signals of selection in the genes of the HLA system and the *SLC22A4* gene, also associated with immune phenotypes, which reinforces the hypothesis of a genetic adaptation of humans to pathogens in the course of the Holocene.

Another element from beyond the grave supports the notion of the evolutionary impact of pathogens on our genomes over the last 10,000 years. It is a 7,000-year-old Mesolithic skeleton. The sequencing of its genome suggests that human adaptation to the pathogens encountered in Europe, through changes in innate-immunity genes, occurred even before the skin of European peoples became pale.

The study of the genomes of dead individuals has also provided us with a better understanding of the way in which the arrival of Europeans in America led to the decline of Native American populations after the introduction of new pathogens. A study conducted in 2016 by the University of Illinois explored the genetic diversity of a population of First Nations – the term used to describe the Native peoples of Canada, before and after the arrival of the Europeans. The genomes of these ancient individuals, who lived in the region of Prince Rupert in British Columbia between 1,000 and 6,000 years ago, were compared to those of individuals of the Tsimshian communities living in the same region today.

First of all, the researchers observed a massive reduction, by about 57%, in the population of the First Nations following their contact with Europeans. Then, after a search for positive selection in the ancient populations, they found that the immune functions appear first, a potential clue to an adaptation to those pathogens present in the ancient environment. Nevertheless, the genomic region that shows the strongest selection signals relates to a gene (*HLA-DQA1*) whose selected mutations are present in all the ancient individuals but in only 36% of the contemporary individuals. The decrease in the frequency of advantageous alleles in the modern population suggests that the arrival of the Europeans brought about environmental changes that made these same mutations deleterious and therefore subject to negative selection in today's First Nations.

All these studies highlight how useful it is to analyse data derived from the DNA of long-dead people. It allows us to measure 'in real time' the action of natural selection and to quantify the way in which selection, whether positive or negative, has participated in shaping the immune system of present-day humans. The dead teach us about our present.

From fossil DNA to tuberculosis

With an aim similar to that of this study of ancient DNA in the First Nations, a study we conducted in collaboration with the team of Jean-Laurent Casanova and Laurent Abel examined the occurrence of negative selection in the context of mycobacterial infections. The tuberculosis bacillus is one of the several mycobacteria, those bacilli that resist acids and alcohol. Tuberculosis is considered among the deadliest infections of our era, with more than 1 billion deaths over the last 2,000 years and still responsible for more than 1.5 million deaths a year. The genetic basis of susceptibility to tuberculosis was little known until the turn of the twenty-first century. It was only in 2018 that the first common mutation in the population associated with an increased risk of developing tuberculosis (a missense mutation in the *TYK2* gene) was identified by Jean-Laurent Casanova's team. This mutation is only observed in populations with European ancestry, at a frequency that varies from 2% to 4% – which nevertheless corresponds to millions of people.

We sought to understand the history of tuberculosis in Europe by studying the trajectory of frequency of this risk-associated mutation, over the last 10,000 years of European history. In looking for the presence of the mutation in more than 1,000 ancient individuals who lived between the Mesolithic period (about 9,000 years ago) and the Middle Ages (within the last 2,000 years), we observed that it appeared for the first time 8,000 years ago in Anatolia and subsequently reached a frequency of almost 10% in Europe towards the end of the Bronze Age, about 3,000 years ago. From the Iron Age, its frequency decreased remarkably to its current 2%–4%. We then asked ourselves the following question: is this decrease in frequency due to the arrival of tuberculosis in Europe, which may have purged the mutation through negative selection, or is it simply due to genetic drift, in other words, to chance? Using specific statistical methods, we were able to show that this decrease in frequency was indeed the result of strong negative selection, which began about 2,000 years ago with one of the strongest selection coefficients in the genome. This study provided genetic evidence of the burden on European health represented by tuberculosis over the last 2,000 years. This is one more illustration of what can be gained from studying ancient DNA: here, the possibility of reconstructing the past of a deadly epidemic about which it is obviously useful to know more, given that it is still very active.

When the loss of a gene proves beneficial

By now, we have seen several examples of new mutations that confer an advantage on their carriers and are thus subject to the effects of positive selection. They often modify the function of a protein, making it more beneficial to human survival. Nevertheless, it sometimes happens that the beneficial trait is not due to the modification of the protein function, but rather to the total loss of a gene.

The loss of a gene (or rather its 'pseudogenization', a pseudogene being the inactive version of a gene following genetic alterations that prevent it from leading to the expression of a protein) has long been considered as systematically associated with deleterious phenotypic consequences. But it may prove beneficial in certain cases and thus be subject to positive selection: this is the 'less-is-more' hypothesis. In the case of immunity, when a pathogen uses immune receptors as a mechanism of entry into the host cell, certain mutations that inactivate these receptors are likely to represent a selective advantage for the host, assuming of course that there are no important pleiotropic effects – in other words, possible collateral damage caused by other functions of the gene.

The case of malaria

The example of the *DARC* (or *ACKR1*) gene, linked to malaria, has become a textbook case in support of the 'less-is-more' hypothesis. *DARC* encodes a co-receptor of *Plasmodium vivax* on the surface of the erythrocytes, that is, the red blood cells, which are the targets of this parasite. When this gene is not expressed, because of a mutation in its promoter, its carrier enjoys almost complete protection against infection by *P. vivax*. The causal mutation is called 'Duffy null'. It is not found in European or Asian populations, whereas its frequency is extremely high (above 50%) in most African populations, reaching near-fixation in Central Africa. This geographical distribution suggests that *P. vivax* has exercised a major selective pressure and would explain the current absence of *P. vivax* malaria in Central Africa. In 2017, a study confirmed the occurrence of positive selection on the Duffy-null mutation, showing that it reached near-fixation in Central Africa starting from a very low frequency (0.1%), with one of the strongest selection advantages ever detected in the human genome. Again, we note the amazing power of genomics to let us see evolution at work.

There is another mutation in the *DARC* gene, completely independent of the one found in Africa, this time among the Papuans. This mutation also decreases the expression of *DARC*. It is associated with a reduced efficiency of *P. vivax* to invade the erythrocytes, which supports the hypothesis that *DARC* deficiency has a protective effect. However, a 2010 study showed that individuals from Madagascar who carry the African Duffy-null mutation and were therefore assumed to be resistant to infection by *P. vivax* were nevertheless infected by this parasite. This suggests that *P. vivax* might also use pathways independent of *DARC* to invade the erythrocytes, casting doubt on the spread and effectiveness of the protection offered by the Duffy-null mutation against *P. vivax*. This observation also questions the nature of the pathogen responsible for the increased prevalence of the Duffy-null mutation in Africa and suggests that there are other pathogens that may also use *DARC* as a way into the host cell.

The case of AIDS

The problem of identifying the true underlying selective agent is equally well illustrated by the *CCR5* gene, which encodes a chemokine receptor that, with another gene (*CD4*), serves as a gateway for HIV, the causal agent of AIDS, to enter human macrophages and monocytes. It has been observed that individuals who are homozygous carriers of mutations leading to a loss of function of this gene (*CCR5*) are strongly protected against infection by HIV. This discovery has had a major impact on research into the virus, particularly as regards the development of new therapeutic strategies. One of these mutations, which makes the CCR5 receptor non-functional, has been much studied. Individuals who are heterozygous carriers of this mutation (named *CCR5-Δ32*, since it leads to the loss of a segment of thirty-two base pairs) display a reduced risk of infection, with a delay in the onset of AIDS, while homozygous carriers are resistant to the infection. This mutation is observed only in Europe, at frequencies that can go as high as 14%, and is almost absent elsewhere.

While the clinical role of the *CCR5-Δ32* mutation is unquestionable, its evolutionary advantage is less clear. On the one hand, some studies have suggested that the mutation has been around for 1,000 years and that it reached its current frequencies through positive selection. On the other hand, these high frequencies could simply be explained by neutral evolution. A study led by Pardis Sabeti's team actually showed that *CCR5-Δ32* has probably evolved under neutrality, as it does not show the expected signatures of strong positive selection in the genome of Europeans.

The controversy over the adaptive nature of this mutation also comes from the date of emergence of the HIV epidemic, which is no more than forty years old. To validate the selective hypothesis, the pathogen or pathogens responsible for the selection pressure on the *CCR5-Δ32* mutation ought therefore to be other pathogens than HIV. As it happens, the date at which this mutation appeared is close to that of the appearance of bubonic plague, which killed about 30% of the population of Europe between 1346 and 1352. This coincidence has led some to advance the hypothesis that the plague bacillus, *Yersinia pestis*, might be the selective agent of the mutation. But it is not the only candidate. The cumulative number of deaths from smallpox over the last 700 years has been much greater than from the plague, since smallpox disproportionately affects young people. A modelling study in population genetics has shown that the smallpox virus may indeed have exerted a selection pressure sufficient to explain the current frequency of the *CCR5-Δ32* mutation. The question remains open.

Noroviruses and gastroenteritis

The noroviruses provide us with a less controversial example of the beneficial effects of the loss of genes. These viruses, which include the Norwalk viruses, are responsible for the majority of gastroenteritis epidemics. Resistance to noroviruses is associated with mutations in a gene (*FUT2*) responsible for the synthesis of the ABO antigens in the epithelial cells and the mucous membranes. Several mutations inactivating this gene – and therefore the secretion of the antigens – have been identified. One nonsense mutation in particular reaches high population frequencies: about 50% in Europe and in Africa and absent in East Asia, with 20% of individuals on average carrying the mutation in a homozygous state, leading to the phenotype called 'non-secretor'. Several studies in population genetics suggest that the high prevalence of this non-secretor phenotype is the result of balancing selection. This is particularly clear in the populations of Europe, the Middle East and sub-Saharan Africa, where we find the secretor phenotype and the non-secretor at intermediate frequencies. The signatures of selection observed may therefore be due, at least partly, to the increased resistance to noroviruses provided by the mutation associated with the non-secretor phenotype.

However, if resistance to noroviruses were the only driving force behind the evolution of the gene in question, the non-secretor phenotype would have increased rapidly in frequency and reached fixation. In fact, in most populations, the two phenotypes are maintained at intermediate frequencies, which suggests that secretor individuals also enjoy a certain advantage. In support of this hypothesis, it appears that secretor individuals have a reduced susceptibility to infections of the urinary tract and to those due to the fungus *Candida* and present increased protection against infection by *Neisseria meningitidis* – meningitis – and streptococci. Here then is an example of balancing selection in which both a mutation of inactivation of a gene and the active version of it are maintained in the general population, the two phenotypes bringing advantages and disadvantages depending on the variable presence, in time and space, of different pathogens.

Total gene loss and septicaemia

A final example worth mentioning is the *CASP12* gene, which is probably the best example of a gene that has been almost completely lost in the human species because of natural selection. The inactive form of this gene, caused by a nonsense mutation that leads to an inactive protein, is virtually fixed outside Africa and presents a frequency of about 80% in African populations. The extremely high frequency of the inactive form seems to be the result of an event of positive selection, which is explained by the increased resistance to septicaemia afforded by this absence. Indeed, individuals carrying the mutation present a stronger inflammatory response to endotoxins: in the case of serious septicaemia, their death rate is three times lower than that of the homozygous carriers of the active form, who present a decreased inflammatory response. This abrogated response may be due to the inhibiting effect of the gene in question (*CASP12*) on another gene (*CASP1*) that participates in the activation of inflammation.

In this context, it is interesting to note the potential evolutionary advantage conferred on chimpanzees by the loss of the gene known as *ICEBERG*, whose product also prevents the activation of the *CASP1* gene and, consequently, the production of inflammatory cytokines. If this advantage is confirmed, it would constitute in its turn an excellent illustration of the way in which humans and chimpanzees, by means of different mechanisms, have followed two independent evolutionary trajectories to confront the risk of septicaemia.

From natural selection to immunological relevance

Be that as it may, examples where the loss of a gene is beneficial remain rare, since in most cases the mutations causing the loss of a gene are deleterious and thus eliminated by negative selection. The more important the function of a gene, the less the organism will tolerate mutations that cause its loss or modification, and the stronger will be the action of negative selection.

For this reason, the detection of genes evolving under strong negative selection is not without interest, given that it may teach us about genes encoding proteins responsible for essential functions and whose variations in activity may lead to serious disorders. The work of my laboratory has shown, for example, that the genes of innate immunity are generally constrained by this selective regime. This is particularly true for genes associated with severe immune disorders in children, like primary immunodeficiencies (also known as 'inborn errors of immunity'). Among the immunity genes evolving under the effect of negative selection, some have been associated by Jean-Laurent Casanova's team with serious disorders, such as herpes simplex encephalitis, pyogenic bacterial infections, Mendelian susceptibility to mycobacterial disease and severe inflammatory diseases. All these examples from clinical genetics clearly highlight how important it is to detect negative selection, allowing us as it does to identify genes of major biological relevance for our survival.

Returning to those mutations that lead to the loss of a gene: with the exception of the examples of beneficial inactivation that have just been mentioned (i.e. the less-is-more hypothesis), in most cases loss-of-function mutations present in the general population are simply due to the redundancy of the genes that carry them. Several population studies at the whole-genome level have looked for the presence of loss-of-function mutations in different human populations around the world. They have shown that every individual carries on average thirty-five mutations in a homozygous state associated with the loss of a gene and that in total we can find about 900 genes displaying mutations leading to a loss of function at a frequency over 1% in at least one of the populations studied. These observations point to genes that have redundant functions and, for mutations at high frequencies in the healthy population, to genes that are unlikely to play a major role in human survival, in other words, genes whose immunological redundancy is almost complete.

The mutations of certain genes that are well known to immunologists support this hypothesis. This is the case of the genes encoding type 1 interferons (IFNs) – a vast group of proteins that help to regulate the activity of the immune system, inducing in particular an antiviral response. Two members of this family (called *IFNA10* and *IFNE*) present nonsense mutations in the general population at high frequencies, which may reach 50% in Asia. In 2011, my team conducted a study in the evolutionary genetics of the human IFNs and showed that the presence of nonsense mutations in *IFNA10* and *IFNE* is compatible with neutrality. This is not the case with other IFNs (like IFN-γ) that show a strong signature of negative selection, which indicates that it plays an essential role. This suggests that the *IFNA10* and *IFNE* genes are largely redundant in the defence of the human host against pathogens.

There is also the example of the gene *TLR5*, which encodes a Toll-like receptor recognizing the bacteria endowed with a flagellum, that is, the filament-like structure allowing them to move about. We find in this gene a nonsense mutation that abolishes the cellular responses to flagellin, a protein that structures the flagella of these bacteria, at relatively high frequencies in most human populations – up to 13% in South-East Asia. In addition, the international gnomAD database (Genome Aggregation Database), comprising genetic data of more than 140,000 individuals from many populations around the world, contains 497 individuals homozygous for this mutation, which indicates that humans can lead a normal life with a complete deficit of *TLR5*, at any rate today. This is consistent with our studies in population genetics, which have shown that the high frequency of this mutation in the general population is compatible with neutrality. These observations also suggest that additional mechanisms for recognizing flagellate bacteria may provide sufficient protection in the absence of *TLR5*.

These examples of genes whose loss is tolerated help us to distinguish, on the one hand, biological functions presenting immunological redundancy from, on the other hand, immunological functions whose loss or alteration may be fatal in the case of attack by infectious agents.

Let's do it together

The examples of natural selection discussed in this chapter, whether they relate to populations alive today or to those that lived in the past, illustrate the adaptability of the genes of the immune system under the effects of positive or balancing selection.

Nevertheless, as seen in the previous chapter, there is a more subtle selection regime, polygenic selection, in which changes in the frequency of mutations of several genes each contribute slightly to adaptation. I have already given a representative example: that of height. The detection of polygenic selection in humans still remains difficult, but it might help us to better understand the genetic basis of many variable phenotypes in our species.

In the case of host–pathogen interactions, as with many other phenotypic traits, the extent of polygenic selection is poorly known. However, the team of Laurent Excoffier developed a method for detecting polygenic selection acting at the level of networks or sub-networks of genes. These studies showed that networks of genes involved in the immune response, or in particular signalling pathways, present clear signatures of polygenic selection. They include genes whose variability is associated with malaria, infection with the bacterium *Escherichia coli*, and interactions between the cytokines and their receptors. The signals of selection detected in these groups of genes are not due to strong signals for a few genes but to small effects on many, which confirms the hypothesis that polygenic selection has been a key mechanism in human adaptation to pathogenic threats.

We still know of few cases of polygenic selection in humans. This does not necessarily mean that they do not exist, it simply means that we do not have at our disposal computational methods with the statistical power necessary to detect them. That is why research into robust statistical methods to detect the extent of polygenic selection in humans is a rapidly growing field in population genetics. Genetics has come a long way – much further than could have been imagined when the molecular structure of DNA was discovered in 1953. But its future is open.

5.

Admixture, Culture and Medicine

Our history as a species, like the history of all living creatures, is recorded in our genes. But human history is not restricted to our genes. Humans also have a cultural past, which sometimes intervenes in their genetic history and inflects it. And it sometimes happens that genetics makes it possible to retrace the history of cultural practices and to answer questions that had previously remained cryptic to human sciences.

In this part, I will examine the way in which genetic admixture and cultural changes have shaped our destiny: genes, languages, peoples and cultures often evolve hand in hand. In this respect, it has to again be underlined that admixture has been a remarkably valuable adaptive resource: our genetic diversity, which is the result of it, is a vital treasure. By retracing the history of migrations and admixture in populations as they are revealed by genetics, genomics once again proves its amazing ability to connect the present with the past, today's immune responses with natural selection in the course of evolutionary times, and also to explain why some mutations once selected for the advantages they provided may, in the new environment that is ours, become burdens. This knowledge, in return, opens up new and promising prospects for medicine.

'Nothing in Biology Makes Sense Except in the Light of Evolution'

This is the title of a 1973 essay by the evolutionary biologist Theodosius Dobzhansky in which he criticizes creationism and defends the theory of evolution. Those opposed to creationism or its more recent avatar, intelligent design, often like to quote this sentence, which takes on its full meaning in the context of evolutionary biology or population genetics: it is by reconstructing what genes tell us about the history of life, by analysing what nature has already accomplished in the course of evolution, that today we are able to understand the mechanisms on which life rests. By studying the past of our species, we gain a greater understanding of its present and may be able to derive lessons that help us to envisage its future.

One thing is certain: we are descended from those of our ancestors who were lucky enough to survive the great epidemics of the past, wars or extreme climatic conditions, who managed to reproduce and pass on to us the genetic resources that had served them so well. What we are today is the fruit of this resilience, which entered our genes through the action of natural selection. Similarly, we cannot ignore the fact that we are the result of a long history of admixture that began at least 60,000 years ago between our ancestors and other hominins, continued throughout our dispersal across the world and during the forced migrations linked, for example, with the slave trade, and persists today between different human populations. How has this very long journey shaped our species? And in particular, how has it shaped our current genetic diversity and our relationship with disease? Genetics, as we have seen, provides tools to give us answers. This knowledge is valuable: by enlightening us about what we are and the journey that has shaped us, it is the bearer of promises for the future, in particular for medicine.

We are all admixed

The genome of each of us is a mosaic composed of the genomes of our ancestors. It is a book of 3 billion letters in which we can read our history. As became clear when the structure of DNA was discovered in 1953, the great book of life is written in the language of genetics. We are gradually learning to decipher it. In it, we find traces of a multitude of encounters that led to our current genetic make-up. These past encounters make us, each one of us – whatever our geographical, ethnic or cultural origins – admixed people. There is no 'pure' lineage: in light of the genetic reality, any claim of identity based on exclusive heritages of 'blood' or race is nothing but fantasy. The concept of race is meaningless from a biological point of view in humans, and 'race' is basically a cultural construction. Our genome is a kind of gigantic jigsaw to which each of our ancestors, with their varied origins, has contributed pieces, and which forms a picture of so many shades, each original and different for every human being. It is a fact: none of us carries a genome with a single ancestral origin.

A few illustrations. In sub-Saharan Africa, all the populations are admixed to different extents. Probably the most significant event of admixture – but far from being the only one – is that associated with the Bantu expansions. The genetic data show clearly that populations speaking Bantu languages admixed with the local populations they encountered throughout their trek across Africa: with the rainforest hunter-gatherers in Central Africa, with the pastoralist peoples of East Africa and with the Khoe-San groups of southern Africa. In their turn, the populations of North Africa are a mosaic of three main components, that coming from the Middle East, that of Maghreb origin and, to a lesser extent, that of sub-Saharan origin. As for Europeans, as we have seen in detail, their genomes reveal that they are the result of admixture events between Mesolithic hunter-gatherers from western Europe, Neolithic farmers from Anatolia and pastoralist nomads from the steppes of Central Asia.

Similar stories, that is, encounters at different periods between different peoples presenting distinct genetic profiles, can also be found in the different regions of Asia and in the Pacific. The Mongol Empire founded by Genghis Khan experienced different expansions after 1206: at its height towards the end of the thirteenth century, it stretched from the Pacific Ocean to the eastern Mediterranean (present-day Turkey). While the cultural effects of the Mongol presence are well documented, its demographic, and therefore genetic, imprint was much less understood until the last few years. But genetic studies have shown that the Mongol expansions did indeed leave a genetic trace in different regions of Eurasia, supporting the idea of admixture events between peoples of Mongol origin and the populations they encountered in Central Asia, the Indian subcontinent, the Middle East and even eastern Europe. Nevertheless, even though the estimated dates of admixture overlap with those of the Mongol Empire, its genetic traces may be confused, at least partly, with those associated with the expansions of the peoples speaking Turkic languages, whose distribution area and cultural influence are very close to those of Genghis Khan's armies.

Even in more circumscribed geographical regions, genetic studies bring to light some surprising histories of admixture. For example, a study conducted by Chris Tyler-Smith of the Wellcome Sanger Institute in the UK focused on the genomic diversity of thirteen individuals who lived between the third and thirteenth centuries in the region corresponding to what is now Lebanon. Nine of them were discovered in a common grave, in Sidon in southern Lebanon, a site that, according to the archeological data, corresponds to a burial pit of Crusaders killed in a battle in the thirteenth century. The genetic data showed that these Crusaders were either of European origin, of local origin, or, in the case of two of them, of mixed

European and Near Eastern origin: evidence that the Crusaders admixed with the local population. However, these events seem to have been brief, since the present-day populations of the region do not present any genetic signature of modern admixture with Europeans. This example illustrates a situation that can probably be generally applied to the whole of human history: the genetic signatures of admixture we observe in today's human populations are merely the tip of the iceberg, in other words, they only reveal *some* of the admixture events that took place – those that have been continuous in time, that is, of a sufficient extent for their traces still to be visible in the present.

It is in the Americas that we find the height of admixture. Practically all the possible genetic components are there: we find the genetic diversity of Native American origin, itself of Asian origin – from the colonial period onwards, and to different extents depending on geographical regions and cultural practices – together with that of European and African origin. Not to mention their respective proportions of Neanderthal ancestry, brought by the Europeans, and Denisovan ancestry, brought by the Asian ancestors of Native Americans. Outside the Americas, there are other populations, admittedly more circumscribed geographically, that are also the result of strong admixture between genetic components coming from far-flung geographical regions. This is the case with the population in South Africa who self-identify as 'Cape Coloureds'. Genetic studies have shown that this population resulted from the admixture between at least five different genetic components over the last 500 years. They derive from the encounter between European colonists, local Bantu-speaking and Khoe-San populations, and populations from the slave trade coming from Bengal (South Asia) and Malaysia (South-East Asia).

Similarly, Madagascar too represents an extreme case of admixture. Even though the island is barely 400 kilometres from the coast of Africa, genetic data of Malagasy populations show that the first inhabitants of the island were Austronesian speakers and came from South-East Asia, probably from southern Borneo: they arrived on

Madagascar in the last 1,500–2,000 years. Then, Bantu populations from southern Africa arrived on the island, and the two populations admixed about 1,000 years ago. To these two major genetic components can be added, to a lesser extent and mainly carried by men, a component of European / Middle Eastern origin, most likely associated with the arrival of groups of Swahili origin or coming from the Arab world. Thus, in the populations of Madagascar, the African component is present on average to a level of 60%, that of Austronesian origin at 36%, and that coming from the Middle East and Europe at 4%, even though the presence of the African and Austronesian components varies a great deal from one individual to another (between 20% and 90%).

Does admixture accelerate adaptation and survival?

Admixture being established as a fact illustrated by genetic data, we can now return to the question that interests us, adapting Dobzhansky's formulation that opened this chapter: What is the meaning of admixture 'in the light of evolution'? Can it be a factor of human adaptation to the environment, and, if so, to what extent? In other words, can admixture be a source of genetic variants conferring greater chances of survival in a given environment? Adaptation is the process by which a population acquires characteristics that allow it to respond more effectively to environmental challenges. It is normally the result of natural selection: in principle it may maintain and spread any new, beneficial mutation. However, theoretical population genetics predicts that a population may also acquire an adaptive variant in a specific environment through genetic exchanges, that is, through admixture with other, closely related populations or species that are already adapted to this environment. The children that are born of this admixture event thereby benefit from the contribution of adaptive genetic elements: a formidable saving of evolutionary time.

In plants and animals

In plants and non-human animals, many empirical observations support the idea that beneficial mutations and adaptive traits are acquired through hybridization with related species, a phenomenon known as *adaptive introgression*. For example, several genomic regions in the sand rock-cress (*Arabidopsis arenosa*), a plant particularly common in north-eastern France, show signals of adaptation to serpentine soils, generally considered poor-quality earth. These genomic regions are the result of introgressions from another plant, the lyre-leaved rock-cress (*Arabidopsis lyrata*). Other examples of adaptive introgression are found in different species of *Heliconius* butterflies, the colour of whose wings comes from other sympatric species; in the domestic mouse (*Mus musculus domesticus*), which has acquired a resistance to rat poison (warfarin in particular) from the Algerian mouse (*Mus spretus*); and in the *Anopheles coluzzii* mosquito, which has acquired resistance to insecticides thanks to hybridization with another mosquito, *Anopheles gambiae*, both also being vectors of malaria.

In humans

In the human species, it is only recently that we have started to study the role of admixture in adaptation to the environment. This phenomenon may occur between different populations, in which case it is called *adaptive admixture*, or between modern humans and archaic humans, that is, through *adaptive introgression*, a process that will be examined later. The multiple episodes of admixture that have marked the history of the African continent, as we have seen, have made today's African populations genomic puzzles derived from multiple ancestries. And it is in Africa that events of adaptive admixture in humans were first observed. In 2012, a genetic study by Chris Tyler-Smith's team of different Ethiopian populations, speaking Semitic and Cushitic languages, showed that they are the result of an admixture event that took place about 3,000 years ago between a genetic component originating in sub-Saharan Africa and a non-African component from the Levant. Migrations from the Levant also introduced into the Ethiopian populations a mutation of a gene associated with a lighter skin pigmentation: it is present at frequencies higher than expected in these populations. This observation supported the action of adaptive admixture, probably as a result of certain social factors such as sexual selection.

Adaptive admixture in the Bantu peoples

Since that first study, others supporting the adaptive role of admixture in humans have accumulated. This is the case with the work of Etienne Patin on the history of the Bantu-speaking populations of Africa. As we have seen, peoples speaking Bantu languages, originally hunter-gatherers living in a region situated between Cameroon and Nigeria, gradually extended their area of habitat. Following a journey of several millennia that began about 4,000 years ago, they spread all over sub-Saharan Africa, where they encountered very diverse ecological habitats: equatorial rainforest, savannah, desert, etc. How did these peoples manage to adapt to such different environments so rapidly? Our study showed that the admixture of the Bantu-speaking peoples with the local populations they encountered allowed them to acquire advantageous genetic variants that facilitated their adaptation to their new habitats. For example, on arriving in East Africa, they inherited from the local pastoralist populations a variant associated with the lactase gene, which makes it possible to digest milk in adulthood. Another important example highlighted by this study is that, thanks to their admixture with the rainforest hunter-gatherers of Central Africa, the Bantu people acquired a new form of the HLA system, contributing to the development of an immune response that was better adapted in case of infection.

Surviving malaria: rainforest hunter-gatherers and the Fulani

The Bantu speakers are not the only winners from these encounters: our work on the HbS mutation of the β-globin gene shows in fact that this admixture was also beneficial to the rainforest hunter-gatherers. This mutation is almost exclusively observed in Africa, where it is present at frequencies that can be as high as 20% in some populations of Central Africa. In the homozygous state, as we have seen, it is responsible for sickle-cell disease, but in the heterozygous state it increases protection against malaria. Where there is no malaria, the mutation ought to disappear in about twelve generations. Hence our argument that dating the mutation should inform us as to the time when malaria started to be a real burden for African populations. We were able to show that it appeared in the ancestors of the Bantu-speaking farmers more than 20,000 years ago, which suggests that these populations were exposed to malaria earlier than originally thought.

On the other hand, the presence of the HbS mutation in the rainforest hunter-gatherers seems to be much more recent. The admixture event through which they acquired it from the Bantu peoples has happened only in the past 6,000 years. Thus, the arrival of farming populations among the forest-dwellers and the deforestation or the manipulation of the environment that ensued over the last 4,000–5,000 years created a fertile ground for the transmission of malaria to the rainforest hunter-gatherers. Following the hypothesis of the so-called 'poison–antidote' model, the Bantu farmers both introduced (unintentionally) to the rainforest hunter-gatherers a new poison, malaria, and transmitted to them at the same time the antidote, the protective HbS mutation, thanks once again to adaptive admixture.

The case of the Fulani is also worthy of mention. This is a traditionally pastoralist population established across the Sahel, from the Atlantic Coast to the Red Sea. A genetic study of the Fulani of Gambia has shown that they are the result of an admixture between Eurasians and populations originally from West Africa that took place about 1,800 years ago. Since then, the Fulani have preserved high levels of Eurasian ancestry in the *LCT* gene, which contains the allele that facilitates the digestion of milk in adults, as well as a high level of African ancestry at the level of the *DARC* gene, whose 'Duffy-null' allele confers resistance to *Plasmodium vivax* malaria. The advantageous nature of the Duffy-null allele seems confirmed by the fact that it is also found at higher frequencies than expected in admixed populations that carry sub-Saharan African ancestry, like in the Sahel, Madagascar, Cabo Verde and Pakistan – where *P. vivax* malaria is endemic.

An 'admixed' immune response

The role of adaptive admixture in the context of the immune response is also supported by certain studies of the HLA system, since some genetic variants of major importance for the establishment of the acquired immune response have been targeted by selection through admixture only a few hundred years ago. We observe, for example, that, in certain populations in Mexico, Puerto Rico and Colombia, the African ancestry is higher than might have been expected at the *HLA* region. We can see in this the effect of a tendency to preserve the *HLA* haplotypes of African origin within these admixed populations, perhaps due to the fact that these haplotypes have been beneficial for the adaptation of the populations to the pathogens encountered in the environment of the New World. In this context, two recent studies have also demonstrated a link between African ancestry and protection against severe dengue in admixed populations in Cuba and Colombia.

All these studies provide evidence that the human adaptation to pathogens may be accelerated by admixture between different populations, although at the moment these observations are often limited to a few cases of populations with African ancestry. As of now, the systematic detection of genomic signatures of adaptive admixture within human populations is convincing proof of the role this process plays in the evolution of traits linked to immunity and provides us with valuable information about the risks of infectious diseases.

Indebted to 'archaic' humans

The study of our genomes shows unequivocally that most individuals and populations on the planet, except those of African origin, are not 100% Sapiens. All individuals from European, Asian, Pacific and Native American populations contain in their genomes between 1% and 5% of genetic material of Neanderthal or Denisovan origin, bearing witness to ancient admixture with these other human forms.

In most cases, however, the introgression of archaic material into the genomes of modern humans was counter-selected, leading to a depletion in archaic inheritance, in particular in the genomic regions encoding genes – the famous 'archaic deserts'. Conversely, the high levels of archaic ancestry that are observed in the genomes of present-day populations indicate either a tolerance of these archaic variants – which may have evolved in a neutral fashion without changing the fitness of modern humans – or the selection of these variants through positive or balancing selection, since they may improve the fitness of modern humans after admixture.

Cases of selection of archaic variants, and thus of adaptive introgression, have been reported for genes linked to morphology, metabolism and responses to environmental factors such as temperature, altitude, sunlight and pathogens. Given that the Neanderthals and the Denisovans lived in Eurasia for at least 300,000 years before the arrival of modern humans, they were probably already well adapted to local foods, as well as to the pathogens and the environmental conditions of the region. So, for modern humans coming from Africa, admixture with the archaic populations they encountered as they entered new regions of the world, about 60,000 years ago, would have had an advantageous effect, facilitating their adaptation to the new environments.

The benefits of Neanderthal DNA

In the last few years, a growing number of studies have endeavoured to identify those genes and biological functions that are subject to adaptive introgression and present contributions from the Neanderthals, in populations outside Africa, and the Denisovans, in the populations of the Pacific, and, to a lesser extent, in Asians. Several functions appear recurrently as targets of adaptive introgression: the morphology and pigmentation of the skin, metabolism, response to hypoxia and immune response. For several genes involved in variation in skin pigmentation and in the differentiation of the keratinocytes and their response to ultraviolet rays (particularly the *BNS2*, *MC1R*, *POU2F3* and *HYAL2* genes), the contribution of Neanderthal variants seems to have been beneficial for the adaptation of human populations to northern latitudes. The Neanderthals also contributed variants linked to the catabolism of lipids, which are found in European populations and may have quickly increased in frequency because of the potential advantages they conferred. Other genes involved in the metabolism of lipids, such as those encoding the transporter proteins SLC16A11 and SLC16A3, have been associated with both an increased risk of developing type 2 diabetes in populations of Latin America and a Neanderthal inheritance. We do not know in what way these variants were adaptive: perhaps they conferred an advantage on Native American populations associated with changes in diet.

Denisovans at altitude

While we have a few hypotheses concerning the contribution of the Neanderthals, the possible adaptive advantages modern humans may have gained from their admixture with the Denisovans remain much less known. The most iconic example is the *EPAS1* gene, associated with the concentration of haemoglobin in the blood and the response to hypoxia. In 2014, Emilia Huerta-Sánchez and Rasmus Nielsen at the University of California in Berkeley showed that this gene displays a high proportion of Denisovan ancestry in the Tibetans, suggesting that the latter received from the Denisovans alleles advantageous for adaptation to life at high altitudes, where the concentration of oxygen is extremely low. The Denisovans also transmitted beneficial variants located in other genes (*TBX15* and *WARS2*) to the ancestors of present-day populations of South and East Asia, Native Americans and Greenlanders. It has been established that the genomic region containing these genes is associated with a large variety of pheno-types ranging from the differentiation of adipose tissue and the distribution of body fat to facial morphology, height, hair pigmen-tation, and the development of the skeleton. These few examples underline the important contribution of the Denisovans to the phenotypic variability of certain present-day human populations.

Archaic humans and immunity

But if there is one biological function that has been affected more than any other by archaic introgression, whether of Neanderthal or Denisovan origin, it has to be the immune response. The first link between archaic introgression and immunity was identified in the *HLA* genomic region, where several haplotypes seem to have been acquired through admixture with the Neanderthals or the Denis-ovans. Since the first study devoted to this question, the list of immunity-related genes for which archaic introgression has proved beneficial to human populations outside Africa has been getting longer and longer.

In 2016, my team showed that the genes involved in the innate immune system, which is the first line of defence against infectious agents, are especially rich in Neanderthal inheritance. This result supports the hypothesis that admixture between Neanderthals and modern humans helped the latter to adapt to the infectious diseases they encountered during the peopling of the planet. For example, advantageous variants of Neanderthal origin, in genes or families of genes involved in the antiviral response (particularly the *STAT2* gene and the *OAS* gene family), are present at high frequencies in most populations outside Africa. Limited to particular populations, a cluster of genes (*TLR1-TLR6-TLR10*), encoding receptors for innate immunity that recognize the microbes on the surface of cells, shows signals of adaptive introgression coming from the Neanderthals, both in Europe and in Asia. The Denisovans, too, made their contribution: it is to them that we owe certain advantageous mutations associated with the immune response in the populations of the Pacific, especially in those of Melanesian ancestry. This is the case, for example, with a gene (*TNFAIP3*) that regulates the immune response and makes it possible to avoid an excessive response that may lead to inflammatory or auto-immune pathologies.

Denisovan defence in the Pacific

It is worth looking more closely at adaptive introgression in the populations of the Pacific. This region is crucial to our understanding of the way interactions with archaic hominins played their part in human adaptation, since the populations of the Pacific present the highest levels of archaic inheritance in the world – up to about 6% if we combine Neanderthal and Denisovan inheritance. A recent study by my team, published in *Nature* in 2021, showed that the Neanderthal contribution facilitated the adaptation of the ancestors of Pacific populations as regards multiple phenotypic traits, such as immunity, neuronal development, metabolism and skin pigmentation. What is surprising is that the adaptive nature of Denisovan introgression is almost exclusively associated with the immune functions! Genes involved in the immune response with a strong Denisovan inheritance include, among others, a gene regulating the immune response mentioned in the previous paragraph (*TNFAIP3*), others involved in immune-cell interactions (*CD33*) or regulating the development of immune cells (*IRF4*). For these examples, the Denisovan variants are found at extremely high frequencies, up to 60%–70%, which suggests that the Denisovans facilitated human adaptation by serving as a reservoir of alleles of resistance to pathogens in the Pacific region.

Archaic variants protecting against viruses

An increasing number of studies show that the archaic variants resulting from introgressions have an impact on human phenotypes at the molecular level, like gene expression. For example, variants of Neanderthal origin impact in particular on the expression of immunity-related genes. This is the case with the family of *OAS* genes, associated with a reduction of the expression of *OAS3* in response to a viral infection. More generally, the mutations associated with the variability of gene expression – they are called expression quantitative trait loci, or eQTLs – are also enriched in Neanderthal inheritance, particularly in the macrophages and the monocytes, cell types that play a major role in the innate immune response.

By way of example, a study by our team revealed that the Neanderthals passed down to Europeans key mutations for the control of the immune response, particularly associated with the response to viral infections, like the influenza virus. The adaptive link between introgression and antiviral response is reinforced by another study, from the team of David Enard and Dmitri Petrov of Stanford University, which shows that the human proteins that interact with viruses (*virus-interacting proteins* or VIPs) are enriched in Neanderthal inheritance, particularly those that interact with RNA viruses like the influenza virus or viruses of the family of *Coronaviridae* (including SARS-CoV-2, responsible for COVID-19!). Overall, these studies underline the adaptive potential of admixture with other hominins like the Neanderthals, and reveal that the latter, already adapted to their infectious environment, facilitated the adaptation of the first Europeans to the selective pressures inflicted by infectious diseases of viral origin.

A double-edged impact on contemporary phenotypes

Beyond molecular phenotypes like gene expression, what is the impact of archaic introgression on macroscopic phenotypes? In other words, even if we observe material of archaic origin in genes associated with different biological functions, in particular immunity, do these archaic segments have a broader effect on modern human phenotypes? Several studies have looked closely at this question, examining to what extent the archaic segments present in our genomes are associated, in medical studies, with variations in human phenotypes, whether these phenotypes are pathological or benign, like certain morphological traits. Several studies have observed overlaps between the presence of archaic variants and different phenotypes, especially those linked to immunity like lupus erythematosus, other auto-immune diseases and allergies.

Two studies have examined this subject systematically by using large human cohorts for which a high number of phenotypes were available. The first, published in *Science* in 2016, evaluated the way in which archaic mutations may have contributed to the variability of more than 1,000 human phenotypes, studying 28,000 individuals of European origin and their medical records. The scientists discovered that variants of Neanderthal origin contribute to the variability of certain human phenotypes of a neurological, psychiatric, immunological and dermatological kind. For example, Neanderthal ancestry appears to be responsible today, at least partly, for an increased risk of skin lesions following exposure to the sun, depression, nicotine addiction and even heart attacks.

The second study, published in 2017 in the *American Journal of Human Genetics*, used the same approach. It looked for correlations between Neanderthal variants and phenotypic traits by analysing the data of about 112,000 individuals from the UK Biobank (the large British cohort for which genetic and phenotypic data have been collected). The scientists observed a link between Neanderthal ancestry and various phenotypes, such as skin and hair pigmentation and sleep rhythms, as well as confirming the links to mood and tobacco addiction. The hypothesis that the introduction of this Neanderthal material confers an evolutionary advantage on the populations in question remains controversial, but these two studies show nonetheless that detecting the introgression of genetic material derived from archaic hominins can teach us about the morphological and physiological variability of today's human populations.

So the Neanderthal heritage that we carry in our genomes – at least those of us of European or Asian origin – is a double-edged sword. On the one hand, it made it possible for our ancestors to adapt better to cold and pathogens, but on the other hand, following the changes in environment and lifestyle that took place over the last 40,000 years, this same heritage may also be deleterious today to our health, as is the case with auto-immune diseases and allergies. And it is in this context that a study that appeared in *Nature* in 2020, conducted by Svante Pääbo's team, identified a link that is surprising, to say the least: a piece of DNA inherited from the Neanderthals may today be an aggravating factor in COVID-19.

Neanderthals and the COVID-19 pandemic

COVID-19, an acute respiratory disease caused by a new coronavirus (SARS-CoV-2), appeared in December 2019 in the city of Wuhan in China and quickly spread around the world. Infections and deaths multiplied: there were almost 4 million deaths around the globe after a year and a half of the pandemic. Since the beginning, it has been clear that old age and being male, as well as other comorbidity factors such as being overweight or having diabetes or hypertension, increase the risk of developing a severe form of COVID-19. However, the strong variability between individuals in their response to the infection, even within the same age group and in particular in older people, strongly suggests that genetic factors also play a key role in the host response to infection by SARS-CoV-2. Indeed, how to explain that, within a group of individuals sharing the same risk factors, some develop relatively benign forms of the disease whereas others develop serious forms, leading in some cases to death?

Very quickly, several studies were launched in an attempt to understand why the individual response to infection by the SARS-CoV-2 virus varies so much from one person to another. The first results derived from genetics came from the Franco-American team of Jean-Laurent Casanova and Laurent Abel. By studying patients afflicted with serious forms of the disease, they showed that about 20% of the severe forms of COVID-19 result from genetic or immunological anomalies that alter the circulating levels of certain proteins playing a major role in the antiviral response (the type 1 IFNs). Since that first discovery, a constantly growing number of studies have examined the genetic architecture of the susceptibility to, or severity of, COVID-19, and several genes or genomic regions have been identified as risk factors. This is particularly the case with a region in chromosome 3 containing six genes, as well as the ABO blood-group system, the genes encoding the antiviral proteins OAS1-OAS2-OAS3 and the *TYK2* gene, the last of these particularly known for being associated with the famous 'cytokine storm': an excessive immune (or hyper-inflammatory) reaction that can cause death.

It is interesting to note that there are also differences between human populations as to the susceptibility to infection by SARS-CoV-2 and the severity of the disease itself. A large multi-ethnic study carried out in the United States has shown that being of Latin American or African American origin is a risk factor. For example, the numbers being hospitalized with severe forms of COVID-19 are very high in African Americans compared with individuals of European ancestry. What is important to underline is that this increased risk persists even after adjustment by other risk factors, such as socio-economic status, age, sex, obesity, type 2 diabetes and hypertension, which suggests that other risk factors, including some genetic factors that are more prevalent in one specific population, may have an impact on the severity of COVID-19.

So far, several independent studies have reached consistent results showing an association between the presence of genetic variants located in chromosome 3 and an increased severity of COVID-19. It is this very region, composed of a block of 50,000 base pairs (a haplotype), that was identified by Svante Pääbo's team as also being of Neanderthal origin – a poison chalice, since carrying this haplotype increases the odds of developing a serious form of the disease by 60%. This genetic risk factor is present in about 16% of Europeans, 50% of Indians and 63% of Bangladeshis. A study conducted independently confirmed the deleterious effects of this haplotype on health, showing that individuals of Bangladeshi origin living in the United Kingdom have twice the risk of succumbing to COVID-19 than the general population.

Are East Asians more 'adapted' to coronaviruses?

A surprising fact regarding this risk factor linked to chromosome 3 is that it is practically absent in the populations of East Asia, like the Chinese, the Japanese and the Vietnamese. Without neglecting the part played by socio-cultural factors, differences in lifestyle or man-agement of the pandemic, such as a stricter lockdown, is it possible that the absence of this risk factor aggravating COVID-19 may explain the lower mortality rate observed, compared with the number of cases, in most East Asian countries? Are Asians better protected biologically? On the one hand, some studies suggest that these differences may be due to immunity naturally acquired by Asians following previous exposures to other coronaviruses for which the main reservoirs are in China, as illustrated by the SARS epidemics starting in 2002 and 2019. On the other hand, it appears that people who have been affected by one of the coronaviruses responsible for the common cold see their risk of dying of COVID-19 diminish by 70%.

The increased exposure of Asian populations to coronaviruses in the past suggests that they are more 'adapted' to respond to these infections. This may partly explain the almost non-existence in this part of the world of the chromosome 3 risk factor, which may have been 'purged' from the population by natural selection. Attempts have been made to test this hypothesis. For example, the team of David Enard, now at the University of Arizona, examined the presence of selection signals at human proteins that interact with coronaviruses (CoV-VIPs) in twenty-six human populations spread over the world. In doing so, they identified a total of forty-two of these proteins that present strong signals of adaptation to the envir-onment and were able to date the beginning of natural selection to about 25,000 years ago. An important detail: the signals of genetic adaptation to coronaviruses (or of natural selection) are only observed in populations originating in East Asia and are absent in all the other populations in the world, including those from South

Asia. In addition, the forty-two CoV-VIPs identified as potential targets of ancient epidemics of coronaviruses (or related viruses) might play a functional role in the aetiology of COVID-19: some of them are currently being used in clinical trials to attenuate the symptoms of COVID-19.

These examples appear as convergent clues: they reinforce the hypothesis that ancient admixtures of Sapiens with other human forms, like the Neanderthals or the Denisovans, played a role as accelerator in the adaptation of our species to the new environmental pressures that it encountered during its journey. But evolution is a constantly moving target where nothing is ever (definitively) acquired – neither strength nor weakness: what was beneficial and facilitated adaptation at the time of admixture, about 40,000 or 50,000 years ago, can be detrimental today. The risk factor on chromosome 3 is a perfect example: inherited from the Neanderthals, its deleterious COVID-19 consequences can only now be understood.

Be that as it may, our evolutionary history may indeed have very tangible effects on our current state of health and the risk of developing certain diseases. What we are today, with our strengths and weaknesses, we owe in part to the exchanges maintained, back in the mists of time, with our distant cousins, archaic humans, whose lineages died out a very long time ago.

Cultural practices and genetic diversity

Throughout this book, we have seen that different factors affect genetic diversity: some create it (mutations), others either favour, eliminate or maintain it in human populations (migration, genetic drift and natural selection). But one of the specificities of our species is its very great cultural diversity. This, just like genes, is transmitted from generation to generation. Cultural traits play a dominant role in human life, including our very survival. They are handed down through the generations to a much greater extent than in other animals. The question of the links between genes and culture is a long-standing one – we often speak of gene–culture co-evolution – but we know more now that our knowledge of the genome has progressed. If there are, in human populations, interactions between cultural differences and biological diversity, is there anything new that genetics can teach us about this? How does culture influence our biological evolution?

Exchanging words, exchanging genes

We know today that some cultural traits specific to our species, such as language, religion and social organization – for example, the caste system in India – limit intermarriages between communities and thus accentuate genetic differences among them. Language is probably the most representative example. To get along, we have to be able to understand each other: we would find it much more difficult to live together if we didn't speak the same language. Historically, it appears that people who spoke the same language were much more likely to marry each other and have children than those who spoke different languages. The work of Luca Cavalli-Sforza in the 1990s established phylogenetic trees based on linguistic distances and others based on genetic distances: by comparing them, he observed a striking concordance. In other words, the genetic proximity between individuals tends to be greater in populations speaking the same language than in populations speaking different languages. This is an example of the way a cultural trait may influence the distribution of genetic diversity between human populations.

The effects of culture on our genetic diversity are not limited to these aspects. Humans have also been able to create new 'ecological niches' after major cultural changes such as the introduction of agriculture. Thus, natural selection may have been modified because of these cultural innovations: some genetic variants are advantageous and make it possible to adapt to this new environment created by humans. A classic example, the adaptation to the digestion of milk in adulthood, is an excellent illustration of this. In populations that practise animal husbandry, milk becomes a major source of nourishment, and the genetic variant making it possible to metabol-ize it becomes advantageous. In this way a particular cultural practice had an impact on the increase in the frequency of this variant.

It is worth examining a few examples to understand how culture and biology interact in the evolution of our species and its relation-ship with diseases.

Migrations and sex

Apart from the correlation between linguistic diversity and genetic diversity, another example showing the impact of cultural practices on genetic diversity should be mentioned, one linked to the difference in migrations depending on sex. Suppose that a group composed exclusively of men undertakes a long migration across a given geo-graphical region and subsequently admixes with women belonging to the different local populations it encounters. The admixed popu-lations resulting from this migration will show a certain homogeneity at the level of their Y chromosomes, since most are descended from the same group of males, while at the level of their mito-chondrial DNA these populations will be quite different, since they are descended from different groups of women. This is an example of the way in which a cultural practice – here differential migration depending on sex – influences the level of genetic variability in a population. What is particularly interesting in population genetics is that we can do the opposite, that is, we analyse the variability of the Y chromosome and mitochondrial DNA to study certain cul-tural practices of a population that are quite distinct depending on sex.

Several studies have attempted this since the end of the 1990s. A pioneering study by Mark Seielstad and Luca Cavalli-Sforza in 1998 showed that, if we compare populations located at a given geograph-ical distance, they tend to show more genetic differences in their Y chromosome than in their mitochondrial DNA. We can interpret these observations as the sign of a higher rate of migration among women than among men, which results in a greater homogeneity in the mitochondrial DNA (which moves with the women) than in the Y chromosomes (derived from the men, who stay where they are). Indeed, even though we know examples of large-scale migrations involving mainly men, like those associated with the armies of Genghis Khan or, to a lesser extent, with the transatlantic slave trade, the genetic data agree on the fact that, historically, women have been more 'mobile' than men. This is a consequence of *residence rules*, which determine the place where a couple settles after marriage. The most common rule in human populations is patrilocal residence: in about 80% of cases, after marriage, the couple will settle in the husband's village. In this way *patrilocality* could explain the more important genetic differences we observe in men (the variations in the Y chromosome, carrier of the paternal history) compared with women (the variations in the mitochondrial DNA, carrier of the maternal history).

This hypothesis was supported in 2001 by a study by Mark Stoneking that compared the paternal and maternal genetic history of groups practising *patrilocality* or *matrilocality*, this latter practice being rare in human populations: matrilocality – where it is the man who leaves home and settles in the wife's village after marriage – is common in fewer than 20% of societies. However, we still find in a region of northern Thailand groups practising both residence rules, which made it possible to test the hypothesis. In perfect agreement with the expected data, within groups practising matrilocality, more differences are observed between individuals at the level of mitochondrial DNA, while, in the groups practising patrilocality, the most marked differences are found at the level of the Y chromosome. Thus, at the level of our species, we can say that women have moved more than men! It is for that reason that, in general, populations tend to resemble one another more from the point of view of mitochondrial DNA than from the point of view of the Y chromosome. Collectively, these studies demonstrate how a cultural choice linked to social organization can influence the distribution of genetic diversity within human populations.

Genetic effects of alliance and descent rules

Alliance rules may also determine the social organization of populations and thereby influence their genetic diversity. There are monogamous societies: they represent about 17% of populations. Others practise different forms of polygyny – that is, the fact that a man may have several wives: they represent between 30% and 80%. As for societies that practise polyandry, where a woman marries several men, they are very rare: less than 1%. For example, in a society where polygyny is generalized, the children will present more resemblances at the level of the Y chromosomes than at the level of mitochondrial DNA. Let us suppose that a man marries ten women and that with each of them he has five children. Let us also imagine that these fifty children are men. In that case, they will all share the same Y chromosome inherited from their father, whereas they will have ten different versions of mitochondrial DNA, inherited from the ten mothers.

Things do not stop there as far as the cultural factors that influence genetic diversity are concerned. Indeed, human societies differ not only through their residence or alliance rules. Other parameters of social organization may also influence the genetic diversity linked to sex. This is the case, for example, with *descent rules*. Descent is the transmission of kinship when one person is descended from another, it defines 'to whom we belong', so to speak. Descent rules define, for example, the transmission of the family name, as is the case in most western societies, but also of the clan or tribe to which an individual belongs in the most traditional societies. There are three great types of descent: *patrilinear*, in which it is the father who transmits kinship (45% of societies), *matrilinear* (12%), where it is the mother who transmits kinship, and *cognatic* (39%), where both parents transmit kinship. In most western societies, such as in Europe, it is cognatic descent that predominates, since we 'belong' to both our father's family and our mother's. What remains to work out is whether descent rules also affect the genetic diversity of populations, and to what extent.

A study conducted in 2004 by Raphaëlle Chaix looked at the role of descent by focusing on a group of populations in Central Asia. These 'traditional' societies are organized into groups of patrilinear descent. An individual belongs to a lineage, the lineages are grouped into clans, and the clans into tribes. According to oral tradition, the members of the same descent group are descended from a common ancestor through the paternal line. The question asked by the scientists was the following: do these descent groups correspond to biological entities or are they simply a social construction? In the hypothesis of a concordance between social and biological groups, two individuals belonging to the same social group (lineage, clan and tribe) should resemble one another more, genetically, than two individuals chosen at random from the population. The genetic kinship of these populations was measured by analysing the Y chromosome, which is also transmitted in a patrilinear way. The results were surprising: whereas individuals belonging to the same lineage or the same clan resemble one another more, genetically speaking, individuals belonging to the same tribe do not resemble one another more than individuals chosen at random from the population. These results demonstrate that lineages and clans do correspond to genuine genetic entities, whereas tribes result from a social, not biological, union of clans of different origins. The common ancestor of a tribe is more a mythical ancestor than a biological ancestor, whose social function may be to strengthen the cohesion of the group.

The caste system in India

The caste system in India, which was practised until very recently, provides us with another example of the influence of cultural factors on genetic diversity. The castes are divisions of societies into groups that are hereditary, hierarchical and often endogamous, marriages within the same caste being strongly encouraged. Occasionally, it might happen that men could marry women of a lower caste, whereas a marriage between a man from a lower caste and a woman from a higher caste was unthinkable! This hierarchical organization led to different movements between the castes depending on sex. As in the case of patrilocality, women seem to have greater mobility: they tend to change caste more often than men. This phenomenon should have had an impact on the distribution of the genetic diversity observed on the one hand at the level of the Y chromosome and on the other hand at the level of mitochondrial DNA.

A study by Michael Bamshad and Lynn Jorde from the University of Utah confirmed these predictions for the first time in 1998. If we compare castes, the genetic differences between them are ten times higher in men (who often stay in the same caste and thus accumulate differences between castes at the level of their Y chromosomes) than in women (who are more mobile and thus homogenize the distribution of mitochondrial DNAs). Once again, these results illustrate the contribution of cultural practices to the shaping of the genetic variability of human populations.

And man got dressed

We know that cultural practices affect genetic diversity. With a little cunning, it then becomes possible to use this same genetic diversity in reverse to detect and identify certain cultural practices from the traces they have left in the genome in the course of human evolution. Here is an example I have chosen for its originality: a study of the appearance of the very first clothes in humans, which tries to date the moment when humans started to dress. What makes this study original is that, to tackle this question, Mark Stoneking's team used not the genetic diversity of humans, but the genetic diversity . . . of lice!

The louse is a parasite that invariably lives alongside humans, and of which we know two varieties, the head louse (*Pediculus humanus capitis*) and the body louse (*Pediculus humanus corporis*). These obligate parasites differ mainly in their habitat: the head louse lives and feeds exclusively on the scalp, whereas the body louse feeds on the body but lives in clothes. Stoneking's team suspected that this ecological specialization probably appeared when humans adopted the frequent use of clothes, an event for which there is no direct archeological evidence. By comparing the genetic diversity of head lice and body lice, they dated their divergence to at least 100,000 years ago in Africa. This result suggests that modern humans started to use the first clothes in Africa in the course of the Middle and Upper Pleistocene. An ingenious use of genetics to trace the history of human cultures!

Agriculture: gene–culture co-evolution

But the cultural practice that has probably most marked the history of our species is the transition from a way of life based on hunting and gathering to a way of life based on agriculture and livestock farming during the Neolithic – a key moment in the progress that shaped our world. This transition, which began about 10,000 years ago and spread from different regions of the globe, led humans to adopt a sedentary lifestyle, bringing about modifications in their chemical, nutritional and infectious environments. For example, the emergence of agriculture and the domestication of animals facilitated the transmission of certain infectious agents, leading to exposure to new zoonoses.

Since the 1990s, a major question has emerged in the field: since we find traces of similar agricultural activities in regions very distant from one another from the Neolithic period onwards, *how* did the new way of life move across the continents? Two opposing models exist, as briefly mentioned in a previous chapter. On the one side, the model of *cultural diffusion* postulates that it was techniques, not people, that travelled: the first farmers simply taught these new practices to the nearby hunter-gatherers, without there being any real population movement. In this case, the diffusion would have taken place without significant admixture between the two groups. On the other side, the *demic diffusion* model supports the idea that agriculture was diffused across the continents as a result of the migrations of farming peoples, who then transmitted to the hunter-gatherers not only new techniques, but also genes, via admixture. Here, too, genetics has shown itself to be a great tool for prehistorians. It has made it possible to distinguish between the two hypotheses, since the genetic data have clearly supported the demic diffusion model. In Europe, the farming peoples coming from the Middle East did indeed leave their genetic traces in most European populations. In Africa, the farming peoples speaking Bantu languages did the same.

Genomic analyses thus confirm that the adoption of agriculture did not occur simply through cultural transmission but that there were also population movements accompanied by admixture. These studies have also supported the idea that the transition to agriculture led to the genetic adaptation of the first farmers in response to the new environmental conditions they faced. For example, a study by my laboratory, conducted in 2016, examined the signals of natural selection on about 1,500 genes involved in the innate immune response in different populations of Africa, Europe and Asia. We identified fifty-seven genes that show strong signals of positive selection in at least one of the populations. Interestingly, we found that most selection events happened in a period from 6,000 to 13,000 years ago, which corresponds to the period when most human populations began to adopt new technologies associated with agriculture. Livestock farming, for example, supposes a proximity with cattle, regular contact with biological waste, and so on. This way of life exposed populations to new pathogens and, as our results suggest, led to phenomena of genetic adaptation of the human host to infectious agents.

Becoming farmers: cause or consequence of expansions?

The dominant hypothesis at the beginning of this century was that, on every continent, agriculture constituted the point of departure for the greatest demographic expansions our species has known. It was thought that the resources generated by agriculture, associated with settlement and the domestication of vegetables and animals, led to an era of abundance and a growth in populations. Once again, genetics would provide us with the facts – and somewhat overturn these received ideas.

It is from Africa that the data have come. For those seeking to understand the way in which the emergence of agriculture impacted on the demographic and adaptive history of humans, Africa provides an excellent model for study. Central Africa, for example, houses the largest number of active hunter-gatherers in the world: the rainforest hunter-gatherers. They cohabit in the same region with their Bantu-speaking neighbours, who are sedentary farmers. It is there that in 2014 we carried out work with Etienne Patin on the genomic variability of more than 300 individuals from both rainforest hunter-gatherers and sedentary farmer populations. Our results somewhat challenged the dominant theory, as far as the African continent is concerned.

According to data derived from archeology and linguistics, the development of agriculture in sub-Saharan Africa began about 5,000 years ago. But our genomic study showed that the main demographic expansion that the ancestors of the farmers experienced was much earlier. Although we cannot rule out the possibility that the first communities of farmers also began to expand demographically 5,000 years ago, our data suggest that in reality the ancestors of the farmers, then hunter-gatherers, experienced a remarkable demographic rise some 7,000–10,000 years ago. This success forced them to adopt a new way of life, to settle and turn to agriculture to meet their growing needs. This study therefore calls into question our ideas on the impact of agriculture in African Neolithic history: it suggests that agriculture was not the direct cause of the demographic success of the populations that adopted it, but rather its consequence.

We also showed that the bulk of admixture between the rainforest hunter-gatherers and the farmers did not begin until about 1,000 years ago. We knew, thanks to the study of their oral traditions and their languages, as well as the genetic diversity of certain microbes that they share, that these populations have been cohabiting and maintaining contact for at least 5,000 years. Yet, this late admixture, which does not fit the classic demographic pattern and testifies to the socio-economic structure peculiar to these populations, was very intense subsequently, to such an extent that today the genomes of some groups of rainforest hunter-gatherers present up to 50% of admixture with the populations of farmers. In addition, this admixture shows a specificity as revealed by genetic analysis – it occurred unilaterally: male farmers admixed with rainforest hunter-gatherer women, but rarely vice versa! As we can see, sex-specific cultural practices have left their traces in the genome of modern human populations.

Hunter-gatherers, farmers and adaptation to pathogens

The ancestors of the rainforest hunter-gatherers and the Bantu-speaking farmers, as we have seen, began to diverge at least 100,000 years ago, which suggests that these two populations have a long history of adaptation to different ecological niches. Selection signals relating to height – a phenotype that seems polygenic and that evolves convergently – have been reported in rainforest hunter-gatherers, as well as functions linked to immunity. In the farmers, most genes that are candidates for adaptation are linked to resistance to malaria, especially in West and Central Africa, which underlines the link between deforestation and increased exposure to the disease. A study conducted by Mattias Jakobsson's team from the University of Uppsala in Sweden explored genetic adaptation to infectious diseases in two groups of Khoe-San hunter-gatherers of South Africa: the ‡Khomani, who had many contacts with the farmers who settled in the region, and the Ju|'hoansi, who remained, historically, more isolated. The researchers noted that immunity genes present strong signals of natural selection in the ‡Khomani, but not in the Ju|'hoansi, suggesting that the adaptation of the immune functions may be rapidly set in motion upon contact with outside groups bringing new pathogens.

Up until very recently, one question remained unanswered: to what extent did the transition from a way of life of mobile hunter-gatherers to a sedentary and essentially agricultural way of life really modify the immune response of these populations? According to the most widespread idea, the emergence of agriculture imposed strong selective pressures on agricultural populations, because of the exposure to new pathogens.

Studies based on the detection of selection signals in these populations do in fact suggest a link between lifestyle and the way in which humans adapt to pathogenic agents in the course of their evolutionary history. But we knew very little about the underlying immunological phenotypes that are behind the molecular signals of selection observed. To learn more, Luis Barreiro's team, from the University of Chicago, examined the Batwa rainforest hunter-gatherers of Uganda and their farmer neighbours, the Bakiga. By combining approaches from population genetics and functional immunology, they demonstrated that the greatest differences in immune responses between these two groups were linked to responses to viral rather than bacterial infections. A detail worth noting is that the selection signals associated with these differences in antiviral response are observed disproportionately in the hunter-gatherers . . . and not in the farmers! This is a surprising result, to say the least, going against the current idea and suggesting that exposure to pathogens – and also, consequently, the processes of genetic adaptation that ensue in the host – is much more dependent on ecological factors (equatorial rainforest versus rural environments) than on the mode of subsistence, such as, in this case, the transition to agriculture. Culture plays its part, of course, but it does not erase the ecological constraints, which remain dominant.

Rural life and urban life: effects on immunity

The tools of genetic analysis open up new prospects for understanding the influence of lifestyle on certain biological parameters, such as immunity. Several studies have looked into the effect of the ecological consequences linked to the lifestyle of populations on the immune response. For example, researchers have examined gene expression in different Arab and Berber communities in southern Morocco, distinguishing them in particular by whether they live in rural or urban environments. These studies have shown that much of the variation in expression of the immunity genes is dictated by urban or rural exposure. They underline the fact that ecological factors associated with place of residence have a major impact on our relationship with infectious agents. True, one might have assumed intuitively that differences in lifestyle and ecological conditions, especially when they are linked to the exposure to pathogens, must have an effect on immune responses and thus modulate the risk of disease. The genomic data confirm this intuition and make it possible to characterize and measure these effects.

Social status and immune response?

So it is now a measurable fact: cultural factors have an effect on the distribution of genetic diversity. This effect may be explored in ever greater detail. For example, we have data concerning the impact of an individual's social status on immune response and health. There can hardly be any doubt that poverty is bad for the health. But biology allows us to learn a little more about this. Health disparities linked to social status may be explained by factors such as access to resources or high-risk behaviours, but there is also evidence of physiological effects of the social environment. It has been shown, for example, that loneliness is associated in humans with high levels of expression of the genes of the inflammatory response. For obvious reasons, experimental studies on humans are excluded. Most studies in this field have therefore turned to the manipulation of social rank in non-human primates.

Jenny Tung's team from Duke University in North Carolina examined the influence of social rank on immune response, using female rhesus macaques as a study model. Macaques live in very hierarchical groups, with well-defined social ranks. This work gave rise to one of the most interesting studies of the last decade. Its publication in *Science* in 2016 showed that social rank has a direct impact on several immune-related phenotypes, such as the proportions of different cell types in the blood, the immune response set in motion by exposure of the cells to lipopolysaccharide (or LPS, an essential component of the bacterial cell wall that mimes bacterial infections) and even the signalling pathways that are used following infection. One particularly interesting fact is that they observed a stronger and much more inflammatory response in females of lower rank, whereas females of higher rank produce a strong anti-viral response. By highlighting the direct biological effects of social inequality on the immune function in great apes, they open up new horizons for studying the potential consequences of social inequality on health in humans.

The underlying mechanisms of these changes in immune response depending on social rank are still unclear, and might be related, in macaques, to stress-induced changes in glucocorticoid metabolism. Such changes trigger epigenetic modifications, altering the accessibility of chromatin and consequently gene expression. In humans, on the other hand, the effects of early childhood environmental conditions, whether nutritional, microbial or psychosocial, on the immune response in adulthood have been mostly assessed (and observed) through the prism of another epigenetic mark, DNA methylation.

As we can see, the epigenetic dimension of the question is of vital importance here. It is therefore necessary to examine it more closely, since this field bridges the gap between the environment and the genome.

Epigenetics: another response to environmental changes

It is impossible to explain the variability of a large number of phenotypic traits by genetic factors alone. In fact, in humans as in various species, in order to respond to environmental cues, the organism has at its disposal means other than genetic adaptation, which was dealt with in Chapter 3. This observation relies on epidemiological studies as well as on countless genetic studies conducted over the last twenty years. An illustrative example is the work conducted on identical twins: even where their genomes are identical, these twins may present different phenotypes, with a different incidence of certain diseases as they age. Large-scale epidemiological data indicate, for example, that the risk of developing cancer depends on genetic factors to a greater or lesser degree depending on the type of cancer. These observations have made it possible to bring back to the forefront the notion of factors external to the genome, factors that are sensitive to the environment and capable of influencing phenotypic variability without modifying the genetic information: the epigenetic mechanisms.

Etymologically, the word epigenetic means 'on top of genetics'. It is used here to describe any molecular modification altering the phenotype that does not involve modification of the actual DNA sequence. Whereas the genome comprises the totality of the material of a cell carrying genetic information (DNA in humans), the epigenome refers to the epigenetic state of the cell, in other words, the totality of the non-genetic actors participating in the regulation of gene expression. We should note that there is very little evidence attesting to the trans-generational inheritance of epigenetic marks in humans. Nevertheless, the diversity of the epigenetic profiles observed in human populations may also explain their phenotypic variations. In this context, epigenetic variations, like the modifications of histones, the action of non-coding RNAs such as the micro-RNAs, or DNA methylation, represent a crucial tool for studying the interface between the environment and the genome.

DNA methylation and phenotypic variability

DNA methylation – in other words, the addition of methyl groups to the DNA – is probably the best-understood component of the epigenetic machinery. It may be altered by environmental factors such as exposure to sunlight, tobacco consumption, diet, stress and exposure to infectious agents. Nevertheless, the variability of DNA methylation may also be affected by genetic factors – mutations in the DNA sequence: these are known as methylation quantitative trait loci (meQTLs). About 20% of the differences between individuals in DNA methylation may be attributed to genetic factors.

Over the last ten years, studies looking for associations between variations in DNA methylation and certain phenotypic traits have multiplied. These studies, at genome-wide level, are called EWAS (epigenome-wide association studies), by analogy with GWAS, of which I have previously spoken. The main factor associated with DNA methylation is age, the profiles of methylation varying considerably over the course of a lifetime. The relationship between methylation and age is so marked that we can use the profiles of DNA methylation to predict the age of an individual with a low rate of error: in general, the difference between biological age and epigenetic age is not much more than one or two years. Other phenotypic traits have been correlated with variations in DNA methylation: some cancers, auto-immune diseases like multiple sclerosis or type I diabetes, body mass index and schizophrenia. However, interpreting the results of EWAS is not as simple as with GWAS, partly because of the confusing effects of the different cell types in the blood (the proportions of different cell types in the blood also influences the methylation profiles), the difficulty of establishing the direction of the link (does the phenotype influence methylation, or vice versa?), as well as the causality of the association (is it dependent on other genetic or epigenetic factors?).

What DNA methylation tells us about populations

Independently of the way DNA methylation affects phenotypic variability, several studies have examined the differences in methylation that may exist between human populations. These studies have highlighted differences in DNA methylation of blood between European and African populations: between 13% and 21% of the sites tested are methylated differently between these populations.

In addition, a comparison of five populations spread throughout the world, allowing a broad coverage of the migratory history of our species, has revealed the importance of genetic factors in explaining the population differences in DNA methylation. Thus, the profiles of DNA methylation make it possible to distinguish broadly different human populations, underlining the possible contribution of epigenetic modifications to phenotypic variation, like physical appearance, drug metabolism, sensory perception and susceptibility to certain diseases. These studies have also revealed that differences in DNA methylation between populations are the result of a combination of different environmental exposures, different allelic frequencies of the genetic variants associated with DNA methylation variation (the meQTLs) and gene–environment interactions.

A recent study conducted in our laboratory focused more specifically on the way in which the differences between human populations affect DNA methylation variation in the context of the immune response. We studied DNA methylation variation at the genome-wide scale in monocytes – white blood cells that play a major role in the innate immune response – and explored the differences between individuals of African ancestry and those of European ancestry. Marked differences in DNA methylation between populations were detected, mainly as regards genes linked to the regulation of immune responses. In particular, about 70% of the sites presenting a differential methylation between Africans and Europeans proved to be associated with a meQTL, confirming the idea that the differences in DNA methylation between human populations are mainly due . . . to genetics itself!

Epigenetics and lifestyle in Africa

Following on from our work on the way in which our species has been able to adapt genetically to the environment, we wanted to understand how changes in habitat and lifestyle, on the one hand, and genetic variability, on the other, affect the DNA methylation profiles of human populations. With this in mind, we focused once again on Central Africa because of the features of human populations inhabiting it: the rainforest hunter-gatherers, mobile groups living in the rainforest, and the Bantu villagers, sedentary farmers in urban, rural or forest habitats. By comparing first the DNA methylation profiles of a group of forest-dwelling Bantu farmers with those of urban or rural Bantu farmers, we observed that the difference in habitat has a significant impact on their epigenome, mainly with regard to the functions of the immune system. On the other hand, by comparing the group of forest-dwelling Bantu farmers with the rainforest hunter-gatherers, who share the same habitat but differ in their historical lifestyle (farmers versus hunter-gatherers), we observed that the main differences at the level of the epigenome were related to functions linked to development, such as height and bone mineral density.

Subsequently, we examined the way in which genetic variability alters the DNA methylation profiles of these populations by identifying meQTLs. We observed that changes in the epigenome associated with recent changes in habitat and affecting immunity are independent of genetic variability, whereas differences in DNA methylation associated with the historical lifestyle of hunter-gatherers and farmers are often driven by genetics and can therefore be inherited. Additionally, and interestingly, these variants controlling DNA methylation profiles often present signals of positive selection. These results suggest that human populations may *respond* initially to environmental variables through epigenetic changes that are not linked to DNA sequence variations, thus attesting to a certain phenotypic plasticity. But, over time, genetic modifications may appear and stabilize the phenotypes, making for a longer-lasting *adaptation* to the environment.

Studying the genomic variability of human populations also teaches us about the biological functions that are the most 'sensitive' to environmental changes, whether they are linked to habitat, lifestyle or exposure to pathogens. For example, the impact that the urbanization of populations has had on the epigenetic profiles of the immune system shows how important it is to examine, as a complement to genetic studies, the way in which epigenetic changes may create an immune-related ground that makes individuals more susceptible to the development of certain immune disorders such as auto-immune, allergic and inflammatory diseases. Genetics, epigenetics and culture are linked to one another, and studying their interactions provides us with valuable information.

Natural selection in the past and immune response today

As we have seen, the action of natural selection exerted by infectious agents on the variability of our genome has been strong and unquestionable. Can we measure its effects? Do these dangerous liaisons maintained in the past between humans and pathogens have consequences on our immune system today? To what extent do human populations present differences in terms of immune response and the risk of developing certain immune or infectious diseases? Finally, what are the factors that contribute to the diversity in immune response between individuals and between human populations? Over the last few years, a growing number of studies in population genetics and systems immunology have tackled these questions, with the aim of gaining a deeper understanding of the extent of the variability of the immune system in humans, the different factors – genetic, evolutionary and environmental – that shape the diversity of the immune response and, finally, the way in which such diversity modifies the risk of developing diseases.

The contribution of studies of gene expression

Although studies in population genetics have allowed us to gain a better understanding of how natural selection has targeted the genes of the immune system, the links between genetic diversity, whether neutral or selected, and immune phenotypes are still poorly understood. Even though many of the differences between individuals, when it comes to the immune response, are due to environmental factors, a fair proportion of this variability is attributable to the genetics of the host. Genome-wide association studies, or GWAS, have identified an increasing number of genetic factors associated with the variability of the immune response, including that leading to a pathological state. Most of these studies have also shown that the genetic variants associated with the risk of disease are often located in regions that regulate gene expression – enhancers in particular. In this context, mapping expression quantitative trait loci (eQTLs) has proved of great value: these variants have provided information that has made it possible to establish links between genetic variability, intermediate phenotypes – such as gene expression – and phenotypes at the level of the organism – such as increased susceptibility to septicaemia, inflammatory bowel disease, viral hepatitis, typhoid fever and tuberculosis.

Epidemiological data show that individuals of different ethnic origins may also differ in their susceptibility to infectious diseases as well as to inflammatory and auto-immune diseases. Recent studies suggest that such disparities may be partly explained by differences in immune response between individuals of different origins. For example, two studies have examined the differences in immune response between individuals of African and European ancestry, measuring in particular the levels of gene expression of immune cells in the presence either of activators of the innate immune response, or of bacteria and viruses (for example, *Mycobacterium tuberculosis*, *Listeria monocytogenes* or the influenza virus). These studies revealed marked differences in response to infection between the two populations. On average, between African and European individuals, 21% of genes seem to present differential expression. This observation may also explain the differences observed between human populations as to their risk of developing certain diseases. In support of this hypothesis, it has been observed that genes that are expressed differently between populations often correspond to genes that have been associated by GWAS with immune-related diseases, such as rheumatoid arthritis, systemic sclerosis and ulcerative colitis, the incidence of which also differs between human populations.

Population differences in inflammatory response

Several studies of the diversity of the human immune response, also supported by epidemiological studies, indicate that individuals of African ancestry present a stronger inflammatory response than individuals of European ancestry. It appears increasingly likely that, setting aside social and environmental effects, a large proportion of these differences can be ascribed to the genetics of the host. Sometimes, a single genetic variant is sufficient to explain such population differences in immune response. For example, in a study published in 2016, my team showed that a single mutation, located in the *TLR1* gene, present in about 70% of Europeans and almost absent in Africans (less than 2%), is behind major differences in inflammatory response between the two populations. In fact, the presence of this mutation leads to a decrease in the expression of about eighty genes linked to the inflammatory response (this mutation being therefore a 'master regulator', or *trans*-eQTL, for its ability to regulate genes located far away on the genome). This observation partly explains the weaker inflammatory response observed in individuals of European ancestry.

The *IRF2* gene has also been identified as a master regulator: a single mutation in this gene has a regulatory effect on hundreds of genes. It is also present at very different frequencies between African and European populations. Consequently, this mutation may be the cause of the differences observed between populations with regard to their response to an important cytokine (IFN-γ) associated with multiple auto-immune and inflammatory disorders. Overall, these studies indicate that the genetic factors of the host may contribute to the differences in immune response observed between human populations.

Natural selection accentuates immune differences

An increasing number of elements suggest not only that the population differences in immune response are partially due to the genetics of the host, but also that natural selection has played an important role in shaping these differences. In fact, among the genes presenting the strongest selection signatures, we observe a major enrichment in those whose expression is different between Africans and Europeans, in particular in macrophages and monocytes. This is an important observation: it constitutes the strongest empirical evidence we have so far that natural selection exerted by infectious agents has crucially influenced the disparities observed between current populations in terms of innate immune response to infections.

Among the examples of mutations that both regulate gene expression and are targeted by natural selection, we find the one that is associated with an attenuated inflammatory response in Europeans (specifically the *trans*-eQTL of *TLR1*). The fact that this mutation presents a strong signal of positive selection in Europeans suggests that a milder inflammatory response has been beneficial to them. More generally, this observation raises questions about an evolutionary conflict between two constraints: on the one hand, developing a strong inflammatory response intended to combat pathogens and, on the other hand, avoiding at the same time the harmful consequences of acute and chronic inflammation, which may lead to tissular lesions and inflammatory and auto-immune diseases.

Collateral damage: auto-immunity, inflammation and obesity

Throughout this book, I have mentioned many examples of genes and biological functions, observing either that their diversity has been preserved for millions of years, or else that, following an environmental change, some mutations have increased in frequency within our species because of the beneficial nature that they conferred when faced with, for example, the presence of pathogens.

Some studies in population genetics, though, have led to a surprising and apparently counter-intuitive observation: the genes whose diversity is correlated with the presence of pathogens are also enriched in genes associated with auto-immune diseases, such as Coeliac disease, ulcerative colitis, type 1 diabetes, Crohn's disease and multiple sclerosis. How could positive selection have then targeted mutations that do not confer protection against diseases, as might have been expected, but on the contrary increase the risk of developing them? The answer to this question is found in the nature of these diseases: they are mostly auto-immune rather than infectious diseases.

Adaptation to pathogens may, in fact, lead to collateral damage in some cases. As Jean-François Bach, an immunologist at the Hôpital Necker in Paris, has shown, the current increase in the frequency of some immune disorders seems to be concomitant with the growing 'sterilization' of modern societies in the course of the twentieth century due to the arrival of antibiotics and vaccines. These epidemiological studies also support the hygiene hypothesis, formulated in 1989 by David P. Strachan of the London School of Hygiene and Tropical Medicine, a hypothesis which postulates that, after an environmental change, some mutations that were advantageous in the past and helped to fight infection are now responsible, in industrialized countries where there is less exposure to infections, for a greater susceptibility to auto-immune, inflammatory and allergic diseases.

A growing number of studies in population genetics also support this hypothesis, showing that positive selection has targeted mutations that increase the risk of certain auto-immune and inflammatory diseases. In addition, the higher frequencies of these risk variants are often observed in populations exposed to strong pathogenic pressure, which suggests that they have played a beneficial role in the defence of the host against infections in the past.

The possible link between infectious diseases and chronic inflammatory disorders is supported by studies showing that some pathogens may be a causal factor in inflammatory and auto-immune diseases: for example, the Epstein–Barr virus and lupus erythematosus, *Mycobacterium avium* complex and Crohn's disease, or *Yersinia enterocolica* and inflammatory bowel disease. Coeliac disease provides us with a great example of this evolutionary mismatch, since mutations in genes such as *IL12A*, *IL18RAP* and *SH2B3* have been targeted by positive selection: individuals carrying these selected alleles, while they enjoy a degree of protection against bacterial infections, present by way of compensation an increased risk of developing Coeliac disease.

More generally, the frequency of variants associated with a number of auto-immune diseases presents a strong differentiation between human populations, further supporting the link between past adaptation and current differences in the risk of disease. Nevertheless, in view of the highly pleiotropic nature of genes involved in immunity – being frequently associated with phenotypes that go beyond response to infection, for example reproduction – it remains difficult to establish a direct link between past selection favouring a specific phenotype and current maladaptation.

This type of evolutionary mismatch is not limited to infectious diseases, or to auto-immune or inflammatory diseases. Epidemiological and genetic data show that such evolutionary mismatches could also have affected the current incidence of certain metabolic disorders, including type 2 diabetes, obesity and gout. As long ago as 1962, the American geneticist James Neel proposed an idea that was revolutionary at the time, and which he called the 'thrifty genotype' hypothesis. According to him, there has been during evolution a selective advantage in favour of those individuals enjoying genes capable of storing food and using it to its best advantage in periods of famine, in other words during most of our history. Following abrupt changes in lifestyle and abundance of food, rich in refined sugars and in fats, these same 'thrifty' genetic variants are today responsible for metabolic havoc in certain populations, explaining the epidemic of diabetes and obesity currently affecting them.

The thrifty genotype hypothesis has been discussed for over sixty years. Yet, despite its popularity, the data to support it remain weak and fragmentary. It is often used to explain the increased frequency of obesity or diabetes in certain populations. For example, the Pima Native Americans (or Akimel O'odham) or the peoples of the South Pacific such as the Polynesians have often been considered archetypes of the thrifty genotype hypothesis 'in action'. However, most genomic studies have failed to support this hypothesis, and very few 'thrifty' genes have been identified convincingly. A study that appeared in 2016 did, nonetheless, identify in the population of Samoa in Polynesia a mutation (in the *CREBRF* gene) associated with a high body mass index and presenting strong indications of positive selection, which suggests that it may be a thrifty variant.

Other factors may also explain the high frequency of metabolic diseases in the Pacific. For example, the same mutations associated with an increased risk of metabolic diseases in the peoples of the Pacific, but which are also found in other populations, may have increased in frequency through simple genetic drift. We know that drift is stronger in small populations, and therefore, because of the decreased efficiency of natural selection, these deleterious mutations may have increased in frequency in the Pacific simply through random processes. In-depth studies, combining historical, cultural, epidemiological, genetic and physiological data, are needed to better understand the reasons for the high incidence of certain metabolic diseases, but also infectious and inflammatory diseases, in different human populations.

Towards a precision medicine

Studies in systems immunology have recently multiplied, aiming to explore the different factors that shape the diversity of the immune system. Most of these studies have examined the individual impact of each of these determinants (genetics, age, sex, etc.) on the immune response, often in the context of a given pathology and on certain types of cells or tissues of interest. Over the past few years, though, several projects have been initiated to produce an integrative analysis of the impact of different factors on the diversity of the immune system. These studies use approaches from systems immunology or population immunology to study the variability of a healthy immune system both in cohorts of twins and in well-defined cohorts of healthy children or adults.

It is in this context that, in 2010, Matthew Albert, an immunologist who at the time was directing a human immunology laboratory at the Institut Pasteur, and myself decided to launch a major study in systems immunology, combining our different but complementary fields of expertise: human immunology on the one hand, human genetics and evolution on the other. We set out to answer several questions: What is a 'healthy' immune response? Can we define the parameters that characterize a healthy immune response in order to identify the thresholds beyond which a breakdown of the immune system leads to a pathology? What are the factors that contribute to the diversity of the immune system in humans? And finally, once these factors have been identified, can we predict the way in which individuals will respond to infection or to a therapeutic treatment? The idea is to use this information to lay the foundations for a precision medicine that would allow us to provide better and more effective treatment. The potential prize is a long way away, but it is not a small one, for patients and for societies, in terms of the quality of care and the cost of treatment.

Our discussions led to the Milieu Intérieur project, launched in 2011. This project, still in progress and co-led today by Darragh Duffy and myself at the Institut Pasteur, brings together the multidisciplinary expertise of researchers in the fields of immunology, infection biology, microbiology, virology, human genetics, bioinformatics, statistics, and evolutionary and systems biology. The project takes its name from the concept developed in the nineteenth century by the physiologist Claude Bernard according to which the *milieu intérieur*, that is, the totality of the internal liquids essential to the lives of animals, governs the very principles of life. It is also nourished by the long tradition of translational research initiated by Louis Pasteur. Its aim is to elucidate the environmental and genetic factors that shape the immune system of a healthy population.

The Milieu Intérieur project relies on the analysis of samples taken from 1,000 healthy donors, stratified by age (five decades, from twenty to sixty-nine) and sex (500 women and 500 men). The tissues sampled comprised fresh peripheral blood, which was stimulated with about thirty pathogens or molecules that activate the innate or acquired immune response, as well as DNA (extracted from the blood), faeces, nasal swabs and skin biopsies. Classic biochemical and haematological tests were carried out, and, for each participant, detailed information was collected about their medical records, lifestyles, diets, tobacco consumption and other parameters that might have an impact on their immune systems. Thirteen years after the launch of this project, we have generated a great deal of data on the variability of the different immune phenotypes and the factors contributing to the latter. These data and analytical approaches may serve as reference for comparative studies with different diseases relating to the immune system.

Two studies from the Milieu Intérieur project strike me as important in better understanding the factors contributing to the diversity of our immune response. The first asked the following question: What are the factors that alter the proportions of the different immune cells in the blood? To answer it, we studied the composition of the peripheral blood in terms of immune cells to evaluate the variability of the immune system in the 1,000 healthy individuals. Some 168 parameters were quantified, including the absolute number and frequency of the immune cells, as well as the levels of expression of the specific markers present on the surface of these cells. By correlating these immune measures with the 140 variables that we obtained for each individual (age, sex, nutrition state, weight, height, vaccination record, variables associated with lifestyle, etc.), we observed that only 5 variables out of the 140 have a real impact on the variability of the immune cells in the blood: the genetics of the host, sex, age, persistent infection by cytomegalo-virus (CMV) and cigarette smoking.

However, these factors do not affect all immune parameters in the same way: while the expression of cell surface markers mainly varies depending on the genetic diversity of the host, especially for cells involved in the innate immune response (monocytes, macro-phages, dendritic cells, etc.), the cellular proportions are more influenced by non-genetic factors in particular for cells involved in the acquired immune response (lymphocytes T and B). Having a longer lifespan, these cells are more likely to be affected by age, tobacco consumption, persistent infections and other non-genetic factors resulting from the individual's way of life.

In the second study, we tackled the same question, but in the con-text of the immune response following infection by pathogens. We analysed the variability of gene expression in the whole blood of the 1,000 individuals, after bacterial stimulations (for example, *Escherichia coli*, *Staphylococcus aureus* and the Calmette–Guérin bacilli or BCG, used as a vaccine against tuberculosis), viral stimula-tions (the influenza virus) and fungal stimulations (*Candida albicans*). The data obtained revealed the following results: about 20% of the variance in the transcriptional response to infection is due to the nat-ural variability of the proportions of different cell types in the blood of each individual; 10% of the variance is explained by the effects of human genetic factors; and, finally, sex and age account for only about 5% of the total variance. Nevertheless, these proportions are only averages across the 572 immune genes studied: the variability of certain genes, like *TLR1*, can be explained above all by the effects of the host genetic factors, while most variability in the expression of other genes, like *STAT1* or *STAT2*, can be explained by age or sex respectively.

However, other factors, including the variation of both epi-genetic marks and the intestinal microbiota, have been reported to impact on immune traits. This makes quantifying the relative contribution of genetic and non-genetic factors – age, sex, diet, persistent viral infections or even socio-economic status – to the variation in immune phenotypes the key to defining the parameters that contribute towards turning a *healthy* immune response into a *pathological* response. Such understanding is crucial if we are to fulfil the promise of a precision medicine that will be able to take the specificity of an individual's biology into account and ensure that public-health strategies take advantage of recent scientific progress.

Epilogue

Let us return to the Gauguin painting we mentioned in the opening pages, and to the three questions it raises. I have tried to show how genomics and population genetics are offering answers to these questions to an unprecedented extent. The genetic study of our past – 'Where do we come from?' – gives us a greater understanding of the sources of our present-day genetic diversity and our relationship to diseases, making it possible to give at least a partial answer to the question: 'What are we?' Is it possible now, at the end of our exploration of these vast spheres of knowledge, to attempt an answer to what is probably the thorniest question of all: 'Where are we going?' Can our knowledge of the past help us to gain a better understanding of how to react to future changes? The answer is yes, at least in part.

Let's look again at the example of infectious diseases. By deciphering the evolutionary history of the immune system and the way in which selection has targeted immunity-related genes in the past, we have learnt a lot about those genes and functions that have played an essential role in our defence against infectious agents. Our immune system is the result of at least 60,000 years of exposure to pathogens and admixture: beneficial immune variants were sometimes acquired by our ancestors thanks to their admixture with archaic humans, then continued to be transmitted through genetic exchanges between different populations of modern humans. There is no reason to suppose that these same variants, genes and functions will not play an important role in dealing with present and future epidemics. So, we can, as of now, orient ourselves towards potential targets that may be used as therapeutic treatments for future infections. But things are not so simple. What was beneficial in the past is not necessarily beneficial today, and will not be beneficial tomorrow: the lifestyles of human populations change, as do their environments. The case of the genomic region of chromosome 3 inherited from Neanderthals and associated today with an increased risk of hospitalization with COVID-19, is a good example, as we have seen. Many links between population genetics and human immunology remain to be explored.

Studying the genetic, evolutionary and environmental factors that make us what we are offers us important avenues to contemplate *where we are going*. Once again, diseases provide us with an excellent example. Genes that evolve under strong negative selection pressures, which do not tolerate (or barely tolerate) mutations in their coding region, are ideal candidates to play an important role in the development of severe disease. It is therefore timely to grant them priority in clinical studies searching the genetic basis of life-threatening diseases. Similarly, the study of the incidence, in human populations, of genetic variants whose clinical impact is important, like those involved in the immune response, has an obvious interest: it may help us to better target populations at risk. For example, a mutation in a type III interferon is associated with spontaneous clearance of the hepatitis C virus and a better response to therapeutic treatment. But the mutation in question presents very variable frequencies depending on human populations: the highest are in Asia, the lowest in Africa. This observation allows us to better distinguish between those populations at risk and those where treatment would be most effective. We have here a concrete example of what a 'precision medicine' might look like, illustrating the way the study of the distribution of genetic diversity in human populations, the knowledge of 'what we are', may have repercussions in the prediction of clinical phenotypes.

Alas, such simple cases are rare. Many other factors come into play when it's a question of predicting the risk of developing a disease or the response to a given treatment. For most phenotypes, the genetic basis is highly complex and polygenic: hundreds of variants may explain the phenotype, as in the case of height or most cancers. These observations have led researchers to propose the use of *polygenic risk scores*, which summarize the effect of many variants on an individual's phenotype. Moreover, genetics is not the only factor driving phenotypic variation. Other parameters, such as sex, age, comorbidities, and other demographic or lifestyle-linked variables, can be added to the genetic factors.

Studies in systems biology (or integrative biology) aim to understand and integrate the different factors that participate in the functioning of biological systems and their diversity. In the long run, they will allow us to, for example, improve our estimates of an individual's risk of developing a disease or predict their response to treatment. It is in this context that studies of cohorts involving thousands of participants, such as the UK Biobank, Biobank Japan, the Estonian Genome Project and the project of the 23andMe biotechnology company, have been launched over the last decade, in order to find associations between a large number of biological and environmental variables on the one hand, and different phenotypes on the other. This is also the case with the previously described Milieu Intérieur project, through which we are gaining a better understanding of how different parameters, including age, sex, genetics, lifestyle and the composition of the microbiota, affect the variability of the human immune response. Once we have accurately identified the way in which the combination of these factors ensures that some of us, for example, present a very reactive immune response whereas in others it is more tempered, we will be able to use this information to *predict* the way in which each individual will respond to infection or to drug therapy.

The aim of these integrative studies is to open the way to a personalized or precision medicine, characterized by a subtle adaptation of the therapeutic treatment to suit the characteristics of the individual. This strategy will make it possible to predict what the most effective medications for each patient will be, while avoiding side effects. For example, in the case of vaccines, it would enable us to optimize the number of doses to suit the immune characteristics of individuals. For the moment, we are only at the beginning. To get there, we will have to acquire much more knowledge about the factors that explain the differences in dealing with disease or treatments in many individuals of varied ethnic, geographical and cultural origins. All these data will then make it possible, using methods based on artificial intelligence such as deep learning, to make predictions at the level of clinical phenotypes. This is a glimpse of the answers to the question 'Where are we going?'

In all these studies – whether involving analysis of genetic association with diseases, estimates of polygenic risk scores, or the establishment of large cohorts to lay the foundations of a precision medicine – one factor remains very much underestimated: the diversity we observe between human populations. What is valid for one population, for example the predictive value of polygenic risk scores, is not necessarily valid for another. In other words, the predictive values of the risk of developing a disease or of having a more efficient response to a given treatment cannot automatically be extrapolated from one human population to another, given their differences: among other factors, the variation of our genetics, the disparity in our environmental exposures and the diversity of our lifestyles.

Unfortunately, most studies in human genomics are conducted on urban populations, mainly of European ancestry. Even though populations of European ancestry represent only 16% of the world's inhabitants, they are the object of more than 78% of genomic studies. For reasons that are both scientific and ethical, these studies should be extended to other populations of non-European ancestry, and in particular to under-represented populations, to better represent the extraordinary range of human diversity. For example, studies in population genetics and medical genetics in populations of African, Native American or Pacific ancestry, among others, having various lifestyles and being exposed to different environments, should provide us with a much more complete view of the biological functions involved in the adaptation of our species to different environments, which would strengthen our knowledge of the links between genetic diversity, natural selection, cultural factors and disease.

Another important question for the future is: How much will our species continue to evolve? Given all our changes in lifestyle, the exploitation of the environment for food production and access to modern medicine, is natural selection still active in our species? The answer is yes. Of course, it takes different forms, but it is still active. Take once again the example of infectious diseases. Towards the end of the nineteenth century, only 35% of the population of Europe reached the age of forty, and only half of children lived to the age of fifteen in the eighteenth century. But, with the improvement in hygienic conditions and the arrival of antibiotics and vaccines, life expectancy has lengthened considerably. That is only the case, however, for those countries having access to modern medicine, which is not true for all of them, especially in Africa. By way of example, at the beginning of this century, 98% of the population reached the age of forty in the United Kingdom, with a life expectancy at birth of around seventy-five to eighty years, whereas in other countries like Mozambique, during the same period, only half the population reached the age of forty.

Even in those regions of the world where humans have been able to *modify* their environment through modern medicine, natural selection continues to act, but in a different way. The arrival of anti-infectious treatments and assisted reproductive technologies has certainly modified the nature and the intensity of natural selection, allowing millions of people to live longer, to reproduce and thus to transmit their genomes. In human populations, because of their slow rate of genetic evolution, the consequences of this situation on the variability of the genome will only be visible in a few thousand years. On the other hand, for microbes, which evolve rapidly, anti-infectious treatments have as of now led to phenomena of adaptation: we often hear of the resistance to antibiotics in bacteria – one of the most serious threats looming over public health today, leading to an increase in mortality, lengthier hospitalizations and increased costs.

The intensity of natural selection in humans has certainly lessened in industrialized countries, as most individuals reach reproductive age, but above all it has changed its target. Whereas in the past selection mainly acted on mortality, it would seem that today it operates more on reproduction. More than 186 million people in the world suffer from infertility: between 8% and 12% of couples of childbearing age are affected by this problem. Among the factors involved in a decrease in male fertility, for example, we could cite obesity, smoking, alcohol and exposure to endocrine disruptors. More striking still, over the last forty years, the number of spermatozoa per millilitre has decreased by more than half. Putting aside the influence of environmental and professional factors and those connected with lifestyle, male infertility and subfertility may have some genetic basis, particularly linked to the Y chromosome. In addition, a study of the population of Denmark, which has yet to be replicated, has shown that certain lineages of the Y chromosome are associated with abnormally low sperm counts, which points to the fact that natural selection is still at work in modern populations.

A 2017 study by Molly Przeworski and Joseph Pickrell tackled the question by directly examining the action of natural selection in today's generations. By analysing the genomes of 60,000 people of European origin from the GERA (Genetic Epidemiology Research on Adult Health and Ageing) cohort and 150,000 people from the UK Biobank, they searched for genetic variants that had changed in frequency between individuals of different ages in order to determine if selection may have acted in one or two generations. Their analyses showed that variants in certain genes are associated with differential survival rates during a lifetime, for example in a gene associated with the risk of developing Alzheimer's (*APOE*) and in another with tobacco addiction (*CHRNA3*). Furthermore, they found an association between a longer lifespan and genetic variants involved in reproductive traits, like the age of puberty and the birth of the first child, which supports the idea that natural selection still operates today.

Independent of its nature or its current intensity, natural selection is only one of the mechanisms through which evolution occurs. It goes without saying that our species is still evolving, since evolution is merely the gradual change of a population within a given environment, which also changes over time. We have seen how admixture between human populations, particularly in Latin America, Africa and Madagascar, is not only a catalyst for human diversity, both genetic and cultural, but also an accelerator of human biological adaptation to the changing environments. It is therefore likely that, given the pace at which environments are being modified today (by, among other factors, climate change, the over-exploitation of natural resources, deforestation, the arrival of new infectious challenges like COVID-19), migrations and increasing admixture between human populations represent key mechanisms of evolution of our species in the near future. Consequently, *Homo sapiens* has not finished evolving . . .

'Diversity, that is my motto,' said Jean de La Fontaine in the seventeenth century. It is also diversity, and diversity alone, that is the driving force behind evolution and the basis of human adaptation to environmental changes. As we have seen, studying the diversity of our genomes makes it possible to answer important questions in anthropology, evolutionary biology and history but also – and of vital importance for the future – in human health. Without diversity, without difference, there is no evolution.

Acknowledgements

This book is the result of twenty-five years of research in human genetics, twenty-five years of questions and, above all, of curiosity. Curiosity about our differences, curiosity about the past of our species and curiosity about the odyssey represented by the search that aims to answer the question: 'What are we?' This led me to population genetics, as an alternative path towards difference, towards the biological and cultural diversity of our species, in space and time.

The study of our evolutionary past gives us a greater understanding not only of *what we are* today, with our similarities and our differences, but also our relationship with diseases. Quite naturally, working at the Institut Pasteur, the temple of microbiology, led me to take an interest in infectious diseases and to study the way in which pathogens have shaped the human genome and our immune response in the course of evolution. Using an approach based on evolutionary genetics, I took an interest in human origins and migrations, the way in which different human populations adapted to a changing environment and the different ways in which our species has been able, with a greater or lesser degree of success, to defend itself against the pathogens responsible for infectious diseases.

I owe a debt to many people and institutions, without whom the journey of writing a book would not have been possible. First of all, I must thank the Institut Pasteur, my adoptive home for the last twenty-five years, for the trust and support it has always given me, just like the CNRS and, more recently, the Collège de France. Thank you also to my mentors, official and unofficial, Silvana Santachiara-Benerecetti, Luca Cavalli-Sforza and Chris Tyler-Smith, models to be followed for their scientific passion, their rigour and their kindness. None of this would have been possible without my colleagues and collaborators. Both scientifically and humanly, I owe a lot to all the members, past or present, of the 'Human Evolutionary Genetics' laboratory, for their passionate work, their patience (especially towards me) and their devotion to collective science. A big thank-you to all my colleagues – I cannot mention them all without taking the risk of forgetting some, but I cannot help but name those to whom I have been particularly close like Laurent Abel, Jean-Laurent Casanova, Antoine Gessain, Evelyne Heyer, Olivier Neyrolles, François Renaud and Paul Verdu. Thank you also to the colleagues who agreed to read or comment on this work during the writing, including Jean-Pierre Changeux, Alain Israel, Etienne Patin, Vinciane Pirenne, Maxime Rotival and Philippe Sansonetti, as well as to Aurélie Bisiaux and Marie-Thérésé Vicente for correcting my French. Finally, this book would not have seen the light of day without the constant support of Jérôme Bon, the editorial help of Marc Kirsch and, last but not least, the patience, talent and always optimistic encouragement of Odile Jacob.

My final thought is for the true protagonists of this work in population genetics, especially the farmers and the hunter-gatherers of Central Africa as well as the different populations of the South Pacific. Without them, none of this would have been possible, and I dedicate this book to them.

Bibliography

1. From Darwin to Genomics

Allison, A. C., 'Protection Afforded by Sickle-Cell Trait against Subtertian Malarial Infection', *British Medical Journal*, 1, 4857 (1954), pp. 290–294.

Anderson, S., Bankier, A. T., Barrell, B. G., de Bruijn, M. H., Coulson, A. R., Drouin, J., Eperon, I. C., *et al.*, 'Sequence and Organization of the Human Mitochondrial Genome', *Nature*, 290, 5806 (1981), pp. 457–465.

Auton, A., Brooks, L. D., Durbin, R. M., Garrison, E. P., Kang, H. M., Korbel, J. O., Marchini, J. L., *et al.*, 'A Global Reference for Human Genetic Variation', *Nature*, 526, 7571 (2015), pp. 68–74.

Bergström, A., McCarthy, S. A., Hui, R., Almarri, M. A., Ayub, Q., Danecek, P., Chen, Y., *et al.*, 'Insights into Human Genetic Variation and Population History from 929 Diverse Genomes', *Science*, 367, eaay5012 (2020).

Bycroft, C., Freeman, C., Petkova, D., Band, G., Elliott, L. T., Sharp, K., Motyer, A., *et al.*, 'The UK Biobank Resource with Deep Phenotyping and Genomic Data', *Nature*, 562, 7726 (2018), pp. 203–209.

Cann, R. L., Stoneking, M., Wilson, A. C., 'Mitochondrial DNA and Human Evolution', *Nature*, 325, 6099 (1987), pp. 31–36.

Cavalli-Sforza, L. L., Menozzi, P., Piazza, A., *The History and Geography of Human Genes*, Princeton, Princeton University Press, 1994.

Chimpanzee Sequencing and Analysis Consortium, 'Initial Sequence of the Chimpanzee Genome and Comparison with the Human Genome', *Nature*, 437, 7055 (2005), pp. 69–87.

Currat, M., Excoffier, L., 'Modern Humans Did Not Admix with Neanderthals during Their Range Expansion into Europe', *PLOS Biology*, 2, 12 (2004), e421.

Darwin, C., *On the Origin of Species by Means of Natural Selection, or the Preservation of Favoured Races in the Struggle for Life*, London, John Murray, 1859.

Darwin, C., *The Descent of Man, and Selection in Relation to Sex*, London, John Murray, 1871.

Enard, W., Przeworski, M., Fisher, S. E., Lai, C. S. L., Wiebe, V., Kitano, T., Monaco, A. P., Pääbo, S., 'Molecular Evolution of *FOXP2*, a Gene Involved in Speech and Language', *Nature*, 418, 6900 (2002), pp. 869–872.

Fisher, R. A., *The Genetical Theory of Natural Selection*, Oxford, Clarendon Press, 1930.

Fu, Q., Li, H., Moorjani, P., Jay, F., Slepchenko, S. M., Bondarev, A. A., Johnson, P. L. F., *et al.*, 'Genome Sequence of a 45,000-Year-Old Modern Human from Western Siberia', *Nature*, 514, 7523 (2014), pp. 445–449.

Fu, W., O'Connor, T. D., Jun, G., Kang, H. M., Abecasis, G., Leal, S. M., Gabriel, S., *et al.*, 'Analysis of 6,515 Exomes Reveals the Recent Origin of Most Human Protein-Coding Variants', *Nature*, 493, 7431 (2013), pp. 216–220.

Hajdinjak, M., Fu, Q., Hübner, A., Petr, M., Mafessoni, F., Grote, S., Skoglund, P., *et al.*, 'Reconstructing the Genetic History of Late Neanderthals', *Nature*, 555, 7698 (2018), pp. 652–656.

Haldane, J. B. S., 'Disease and Evolution', *Rice Science*, 19 (Suppl. A) (1949), pp. 68–76.

Hart, D. L., Clark, A. G., *Principles of Population Genetics*, Sunderland, Sinauer Associates, Inc. Publishers, 2007.

Higuchi, R., Bowman, B., Freiberger, M., Ryder, O. A., Wilson, A. C., 'DNA Sequences from the Quagga, an Extinct Member of the Horse Family', *Nature*, 312, 5991 (1984), pp. 282–284.

International Human Genome Sequencing Consortium, 'Finishing the Euchromatic Sequence of the Human Genome', *Nature*, 431, 7011 (2004), pp. 931–945.

Johanson, D. C., White, T. D., Coppens, Y., 'A New Species of the Genus *Australopithecus* (Primates: Hominidae) from the Pliocene of Eastern Africa', *Kirtlandia*, 28 (1978), pp. 1–14.

Kimura, M., 'Evolutionary Rate at the Molecular Level', *Nature*, 217, 5129 (1968), pp. 624–626.

Kimura, M., *The Neutral Theory of Molecular Evolution*, Cambridge, Cambridge University Press, 1984.

Krings, M., Stone, A., Schmitz, R. W., Krainitzki, H., Stoneking, M., Pääbo, S., 'Neandertal DNA Sequences and the Origin of Modern Humans', *Cell*, 90, 1 (1997), pp. 19–30.

Kuderna, L. F., Esteller-Cucala, P., Marques-Bonet, T., 'Branching Out: What Omics Can Tell Us about Primate Evolution', *Current Opinion in Genetics & Development*, 62 (2020), pp. 65–71.

Kuhlwilm, M., de Manuel, M., Nater, A., Greminger, M. P., Krützen, M., Marques-Bonet, T., 'Evolution and Demography of the Great Apes', *Current Opinion in Genetics & Development*, 41 (2016), pp. 124–129.

Lander, E. S., Linton, L. M., Birren, B., Nusbaum, C., Zody, M. C., Baldwin, J., Devon, K., *et al.*, 'Initial Sequencing and Analysis of the Human Genome', *Nature*, 409, 6822 (2001), pp. 860–921.

Landsteiner, K., 'Über Agglutinationserscheinungen Normalen Menschlichen Blutes', *Wiener klinische Wochenschrift*, 14 (1901), pp. 1132–1134.

Lek, M., Karczewski, K. J., Minikel, E. V., Samocha, K. E., Banks, E., Fennell, T., O'Donnell-Luria, A. H., *et al.*, 'Analysis of Protein-Coding Genetic Variation in 60,706 Humans', *Nature*, 536, 7616 (2016), pp. 285–291.

Lewontin, R. C., 'The Apportionment of Human Diversity', in T. Dobzhansky, M. K. Hecht and W. C. Steere (eds.), *Evolutionary Biology*, New York, Appleton-Century-Crofts, 1972, pp. 381–398.

Locke, D. P., Hillier, L. W., Warren, W. C., Worley, K. C., Nazareth, L. V., Muzny, D. M., Yang, S. P., *et al.*, 'Comparative and Demographic Analysis of Orang-Utan Genomes', *Nature*, 469, 7331 (2011), pp. 529–533.

Mafessoni, F., Grote, S., de Filippo, C., Slon, V., Kolobova, K. A., Viola, B., Markin, S. V., *et al.*, 'A High-Coverage Neandertal Genome from Chagyrskaya Cave', *Proceedings of the National Academy of Sciences USA*, 117, 26 (2020), pp. 15132–15136.

Maixner, F., Krause-Kyora, B., Turaev, D., Herbig, A., Hoopmann, M. R., Hallows, J. L., Kusebauch, U., *et al.*, 'The 5,300-Year-Old *Helicobacter Pylori* Genome of the Iceman', *Science*, 351, 6269 (2016), pp. 162–165.

Mallick, S., Li, H., Lipson, M., Mathieson, I., Gymrek, M., Racimo, F., Zhao, M., *et al.*, 'The Simons Genome Diversity Project: 300 Genomes from 142 Diverse Populations', *Nature*, 538, 7624 (2016), pp. 201–206.

Marques-Bonet, T., Ryder, O. A., Eichler, E. E., 'Sequencing Primate Genomes: What Have We Learned?', *Annual Review of Genomics and Human Genetics*, 10 (2009), pp. 355–386.

Meyer, M., Kircher, M., Gansauge, M. T., Li, H., Racimo, F., Mallick, S., Schraiber, J. G., *et al.*, 'A High-Coverage Genome Sequence from an Archaic Denisovan Individual', *Science*, 338, 6104 (2012), pp. 222–226.

Perry, G. H., Yang, F., Marques-Bonet, T., Murphy, C., Fitzgerald, T., Lee, A. S., Hyland, C., *et al.*, 'Copy Number Variation and Evolution in Humans and Chimpanzees', *Genome Research*, 18, 11 (2008), pp. 1698–1710.

Petr, M., Hajdinjak, M., Fu, Q., Essel, E., Rougier, H., Crevecoeur, I., Semal, P., *et al.*, 'The Evolutionary History of Neanderthal and Denisovan Y Chromosomes', *Science*, 369, 6511 (2020), pp. 1653–1656.

Prüfer, K., de Filippo, C., Grote, S., Mafessoni, F., Korlević, P., Hajdinjak, M., Vernot, B., *et al.*, 'A High-Coverage Neandertal Genome from Vindija Cave in Croatia', *Science*, 358, 6363 (2017), pp. 655–658.

Prüfer, K., Racimo, F., Patterson, N., Jay, F., Sankararaman, S., Sawyer, S., Heinze, A., *et al.*, 'The Complete Genome Sequence of a Neanderthal from the Altai Mountains', *Nature*, 505, 7481 (2014), pp. 43–49.

Quintana-Murci, L., Semino, O., Bandelt, H.-J., Passarino, G., McElreavey, K., Santachiara-Benerecetti, A. S., 'Genetic Evidence of an Early Exit of *Homo Sapiens Sapiens* from Africa through Eastern Africa', *Nature Genetics*, 23, 4 (1999), pp. 437–441.

Rasmussen, M., Li, Y., Lindgreen, S., Pedersen, J. S., Albrechtsen, A., Moltke, I., Metspalu, M., *et al.*, 'Ancient Human Genome Sequence of an Extinct Palaeo-Eskimo', *Nature*, 463, 7282 (2010), pp. 757–762.

Reich, D., Green, R. E., Kircher, M., Krause, J., Patterson, N., Durand, E. Y., Viola, B., *et al.*, 'Genetic History of an Archaic Hominin Group from Denisova Cave in Siberia', *Nature*, 468, 7327 (2010), pp. 1053–1060.

Sikora, M., Carpenter, M. L., Moreno-Estrada, A., Henn, B. M., Underhill, P. A., Sanchez-Quinto, F., Zara, I., *et al.*, 'Population Genomic Analysis of Ancient and Modern Genomes Yields New Insights into the Genetic Ancestry of the Tyrolean Iceman and the Genetic Structure of Europe', *PLOS Genetics*, 10, 5 (2014), e1004353.

Skoglund, P., Mathieson, I., 'Ancient Genomics of Modern Humans: The First Decade', *Annual Review of Genomics and Human Genetics*, 19 (2018), pp. 381–404.

Slon, V., Mafessoni, F., Vernot, B., de Filippo, C., Grote, S., Viola, B., Hajdinjak, M., *et al.*, 'The Genome of the Offspring of a Neanderthal Mother and a Denisovan Father', *Nature*, 561, 7721 (2018), pp. 113–116.

Watson, J. D., Crick, F. H., 'Molecular Structure of Nucleic Acids: A Structure for Deoxyribose Nucleic Acid', *Nature*, 171, 4356 (1953), pp. 737–738.

Wright, S., *Evolution and the Genetics of Populations: The Theory of Gene Frequencies*, Chicago, University of Chicago Press, 1969.

Yang, M. A., Fu, Q., 'Insights into Modern Human Prehistory Using Ancient Genomes', *Trends in Genetics*, 34, 3 (2018), pp. 184–196.

2. *Journeys and Encounters*

Adhikari, K., Chacón-Duque, J. C., Mendoza-Revilla, J., Fuentes-Guajardo, M., Ruiz-Linares, A., 'The Genetic Diversity of the Americas', *Annual Review of Genomics and Human Genetics*, 18 (2017), pp. 277–296.

Bae, C. J., Douka, K., Petraglia, M. D., 'On the Origin of Modern Humans: Asian Perspectives', *Science*, 358, 6368 (2017).

Behar, D. M., Harmant, C., Manry, J., Van Oven, M., Haak, W., Martinez-Cruz, B., Salaberria, J., *et al.*, 'The Basque Paradigm: Genetic Evidence of a Maternal Continuity in the Franco-Cantabrian Region since Pre-Neolithic Times', *American Journal of Human Genetics*, 90, 3 (2012), pp. 486–493.

Belbin, G. M., Nieves-Colón, M. A., Kenny, E. E., Moreno-Estrada, A., Gignoux, C. R., 'Genetic Diversity in Populations across Latin America: Implications for Population and Medical Genetic Studies', *Current Opinion in Genetics & Development*, 53 (2018), pp. 98–104.

Berniell-Lee, G., Calafell, F., Bosch, E., Heyer, E., Sica, L., Mouguiama-Daouda, P., Van der Veen, L., *et al.*, 'Genetic and Demographic Implications of the Bantu Expansion: Insights from Human Paternal Lineages', *Molecular Biology and Evolution*, 26, 7 (2009), pp. 1581–1589.

Blum, M. G., Jakobsson, M., 'Deep Divergences of Human Gene Trees and Models of Human Origins', *Molecular Biology and Evolution*, 28, 2 (2011), pp. 889–898.

Brauer, G., 'The Evolution of Modern Humans: A Comparison of the African and Non-African Evidence', in P. Mellars and C. B. Stringer (eds.), *The Human Revolution: Behavioural and Biological Perspectives on the Origins of Modern Humans*, Edinburgh, Edinburgh University Press, 1989, pp. 123–154.

Browning, S. R., Browning, B. L., Zhou, Y., Tucci, S., Akey, J. M., 'Analysis of Human Sequence Data Reveals Two Pulses of Archaic Denisovan Admixture', *Cell*, 173, 1 (2018), pp. 53–61, e9.

Brunet, M., Guy, F., Pilbeam, D., Mackaye, H. T., Likius, A., Ahounta, D., Beauvilain, A., *et al.*, 'A New Hominid from the Upper Miocene of Chad, Central Africa', *Nature*, 418, 6894 (2002), pp. 145–151.

Bryc, K., Auton, A., Nelson, M. R., Oksenberg, J. R., Hauser, S. L., Williams, S., Froment, A., *et al.*, 'Genome-Wide Patterns of Population Structure and Admixture in West Africans and African Americans', *Proceedings of the National Academy of Sciences USA*, 107, 2 (2010), pp. 786–791.

Campbell, M. C., Hirbo, J. B., Townsend, J. P., Tishkoff, S. A., 'The Peopling of the African Continent and the Diaspora into the New World', *Current Opinion in Genetics & Development*, 29 (2014), pp. 120–132.

Campbell, M. C., Tishkoff, S. A., 'African Genetic Diversity: Implications for Human Demographic History, Modern Human Origins, and Complex Disease Mapping', *Annual Review of Genomics and Human Genetics*, 9 (2008), pp. 403–433.

Cann, R. L., Stoneking, M., Wilson, A. C., 'Mitochondrial DNA and Human Evolution', *Nature*, 325, 6099 (1987), pp. 31–36.

Choin, J., Mendoza-Revilla, J., Arauna, L. R., Cuadros-Espinoza, S., Cassar, O., Larena, M., Ko, A. M., *et al.*, 'Genomic Insights into Population History and Biological Adaptation in Oceania', *Nature*, 592, 7855 (2021), pp. 583–589.

de Filippo, C., Barbieri, C., Whitten, M., Mpoloka, S. W., Gunnarsdóttir, E. D., Bostoen, K., Nyambe, T., *et al.*, 'Y-Chromosomal Variation in Sub-Saharan Africa: Insights into the History of Niger-Congo Groups', *Molecular Biology and Evolution*, 28, 3 (2011), pp. 1255–1269.

de Filippo, C., Bostoen, K., Stoneking, M., Pakendorf, B., 'Bringing Together Linguistic and Genetic Evidence to Test the Bantu Expansion', *Proceedings of the Royal Society B: Biological Sciences*, 279, 1741 (2012), pp. 3256–3263.

Ehret, C., 'Bantu Expansions: Re-Envisioning a Central Problem of Early African History', *International Journal of African Historical Studies*, 34, 1 (2001), pp. 5–41.

Fu, Q., Hajdinjak, M., Moldovan, O. T., Constantin, S., Mallick, S., Skoglund, P., Patterson, N., *et al.*, 'An Early Modern Human from Romania with a Recent Neanderthal Ancestor', *Nature*, 524, 7564 (2015), pp. 216–219.

Fu, Q., Meyer, M., Gao, X., Stenzel, U., Burbano, H. A., Kelso, J., Pääbo, S., 'DNA Analysis of an Early Modern Human from Tianyuan Cave, China', *Proceedings of the National Academy of Sciences USA*, 110, 6 (2013), pp. 2223–2227.

GenomeAsia100K Consortium, 'The GenomeAsia 100k Project Enables Genetic Discoveries across Asia', *Nature*, 576, 7785 (2019), pp. 106–111.

Gomez, F., Hirbo, J., Tishkoff, S. A., 'Genetic Variation and Adaptation in Africa: Implications for Human Evolution and Disease', *Cold Spring Harbor Perspectives in Biology*, 6, 7 (2014), a008524.

Gosling, A. L., Matisoo-Smith, E. A., 'The Evolutionary History and Human Settlement of Australia and the Pacific', *Current Opinion in Genetics & Development*, 53 (2018), pp. 53–59.

Gravel, S., Zakharia, F., Moreno-Estrada, A., Byrnes, J. K., Muzzio, M., Rodriguez-Flores, J. L., Kenny, E. E., *et al.*, 'Reconstructing Native American Migrations from Whole-Genome and Whole-Exome Data', *PLOS Genetics*, 9, 12 (2013), e1004023.

Greenberg, J. H., 'Linguistic Evidence Regarding Bantu Origins', *Journal of African History*, 13 (1972), pp. 189–216.

Grollemund, R., Branford, S., Bostoen, K., Meade, A., Venditti, C., Pagel, M., 'Bantu Expansion Shows that Habitat Alters the Route and Pace of Human Dispersals', *Proceedings of the National Academy of Sciences USA*, 112, 43 (2015), pp. 13296–13301.

Haak, W., Lazaridis, I., Patterson, N., Rohland, N., Mallick, S., Llamas, B., Brandt, G., *et al.*, 'Massive Migration from the Steppe Was a Source for Indo-European Languages in Europe', *Nature*, 522, 7555 (2015), pp. 207–211.

Hammer, M. F., Woerner, A. E., Mendez, F. L., Watkins, J. C., Wall, J. D., 'Genetic Evidence for Archaic Admixture in Africa', *Proceedings of the National Academy of Sciences USA*, 108, 37 (2011), pp. 15123–15128.

Harris, K., Nielsen, R., 'The Genetic Cost of Neanderthal Introgression', *Genetics*, 203, 2 (2016), pp. 881–891.

Hellenthal, G., Busby, G. B. J., Band, G., Wilson, J. F., Capelli, C., Falush, D., Myers, S., 'A Genetic Atlas of Human Admixture History', *Science*, 343, 6172 (2014), pp. 747–751.

Henn, B. M., Botigue, L. R., Gravel, S., Wang, W., Brisbin, A., Byrnes, J. K., Fadhlaoui-Zid, K., *et al.*, 'Genomic Ancestry of North Africans Supports Back-to-Africa Migrations', *PLOS Genetics*, 8, 1 (January 2012), e1002397.

Henn, B. M., Cavalli-Sforza, L. L., Feldman, M. W., 'The Great Human Expansion', *Proceedings of the National Academy of Sciences USA*, 109, 44 (2012), pp. 17758–17764.

Henn, B. M., Steele, T. E., Weaver, T. D., 'Clarifying Distinct Models of Modern Human Origins in Africa', *Current Opinion in Genetics & Development*, 53 (2018), pp. 48–56.

Hershkovitz, I., Weber, G. W., Quam, R., Duval, M., Grün, R., Kinsley, L., Ayalon, A., *et al.*, 'The Earliest Modern Humans Outside Africa', *Science*, 359, 6374 (2018), pp. 456–459.

Holden, C. J., 'Bantu Language Trees Reflect the Spread of Farming across Sub-Saharan Africa: A Maximum-Parsimony Analysis', *Proceedings of the Royal Society B: Biological Sciences*, 269, 1493 (2002), pp. 793–799.

Hsieh, P., Veeramah, K. R., Lachance, J., Tishkoff, S. A., Wall, J. D., Hammer, M. F., Gutenkunst, R. N., 'Whole-Genome Sequence Analyses of Western Central African Pygmy Hunter-Gatherers Reveal a Complex Demographic History and Identify Candidate Genes under Positive Natural Selection', *Genome Research*, 26, 3 (2016), pp. 279–290.

Hsieh, P., Woerner, A. E., Wall, J. D., Lachance, J., Tishkoff, S. A., Gutenkunst, R. N., Hammer, M. F., 'Model-Based Analyses of Whole-Genome Data Reveal a Complex Evolutionary History Involving Archaic Introgression in Central African Pygmies', *Genome Research*, 26, 3 (2016), pp. 291–300.

Hudjashov, G., Karafet, T. M., Lawson, D. J., Downey, S., Savina, O., Sudoyo, H., Lansing, J. S., Hammer, M. F., Cox, M. P., 'Complex Patterns of Admixture across the Indonesian Archipelago', *Molecular Biology and Evolution*, 34, 10 (2017), pp. 2439–2452.

Ingman, M., Kaessmann, H., Pääbo, S., Gyllensten, U., 'Mitochondrial Genome Variation and the Origin of Modern Humans', *Nature*, 408, 6813 (2000), pp. 708–713.

Ioannidis, A. G., Blanco-Portillo, J., Sandoval, K., Hagelberg, E., Miquel-Poblete, J. F., Moreno-Mayar, J. V., Rodríguez-Rodríguez, J. E., *et al.*, 'Native American Gene Flow into Polynesia Pre-Dating Easter Island Settlement', *Nature*, 583, 7817 (2020), pp. 572–577.

Jacobs, G. S., Hudjashov, G., Saag, L., Kusuma, P., Darusallam, C. C., Lawson, D. J., Mondal, M., *et al.*, 'Multiple Deeply Divergent Denisovan Ancestries in Papuans', *Cell*, 177, 4 (2019), pp. 1010–1021, e32.

Jakobsson, M., Scholz, S. W., Scheet, P., Gibbs, J. R., VanLiere, J. M., Fung, H.-C., Szpiech, Z. A., *et al.*, 'Genotype, Haplotype and Copy-Number Variation in Worldwide Human Populations', *Nature*, 451, 7181 (2008), pp. 998–1003.

Jarvis, J. P., Scheinfeldt, L. B., Soi, S., Lambert, C., Omberg, L., Ferwerda, B., Froment, A., *et al.*, 'Patterns of Ancestry, Signatures of Natural Selection, and Genetic Association with Stature in Western African Pygmies', *PLOS Genetics*, 8, 4 (2012), e1002641.

Jeong, C., Balanovsky, O., Lukianova, E., Kahbatkyzy, N., Flegontov, P., Zaporozhchenko, V., Immel, A., *et al.*, 'The Genetic History of Admixture across Inner Eurasia', *Nature Ecology & Evolution*, 3, 6 (2019), pp. 966–976.

Jobling, M. A., Hollox, E., Kivisild, T., Tyler-Smith, C., *Human Evolutionary Genetics*, New York, Taylor & Francis, Garland Science, 2013.

Lachance, J., Vernot, B., Elbers, C. C., Ferwerda, B., Froment, A., Bodo, J. M., Lema, G., *et al.*, 'Evolutionary History and Adaptation from High-Coverage Whole-Genome Sequences of Diverse African Hunter-Gatherers', *Cell*, 150, 3 (2012), pp. 457–469.

Laval, G., Patin, E., Barreiro, L. B., Quintana-Murci, L., 'Formulating a Historical and Demographic Model of Recent Human Evolution Based on Resequencing Data from Noncoding Regions', *PLOS One*, 5, 4 (2010), e10284.

Lazaridis, I., 'The Evolutionary History of Human Populations in Europe', *Current Opinion in Genetics & Development*, 53 (2018), pp. 21–27.

Lazaridis, I., Patterson, N., Mittnik, A., Renaud, G., Mallick, S., Kirsanow, K., Sudmant, P. H., *et al.*, 'Ancient Human Genomes Suggest Three Ancestral Populations for Present-Day Europeans', *Nature*, 513, 7518 (2014), pp. 409–413.

Leslie, S., Winney, B., Hellenthal, G., Davison, D., Boumertit, A., Day, T., Hutnik, K., *et al.*, 'The Fine-Scale Genetic Structure of the British Population', *Nature*, 519, 7543 (2015), pp. 309–314.

Li, S., Schlebusch, C., Jakobsson, M., 'Genetic Variation Reveals Large-Scale Population Expansion and Migration during the Expansion of Bantu-Speaking Peoples', *Proceedings of the Royal Society B: Biological Sciences*, 281, 20141448 (2014).

Lipson, M., Cheronet, O., Mallick, S., Rohland, N., Oxenham, M., Pietrusewsky, M., Pryce, T. O., *et al.*, 'Ancient Genomes Document Multiple Waves of Migration in Southeast Asian Prehistory', *Science*, 361, 6397 (2018), pp. 92–95.

Lipson, M., Loh, P.-R., Patterson, N., Moorjani, P., Ko, Y.-C., Stoneking, M., Berger, B., Reich, D., 'Reconstructing Austronesian Population History in Island Southeast Asia', *Nature Communications*, 5 (2014), p. 4689.

Lipson, M., Skoglund, P., Spriggs, M., Valentin, F., Bedford, S., Shing, R., Buckley, H., *et al.*, 'Population Turnover in Remote Oceania Shortly after Initial Settlement', *Current Biology*, 28, 7 (2018), pp. 1157–1165, e7.

Malaspinas, A.-S., Westaway, M. C., Muller, C., Sousa, V. C., Lao, O., Alves, I., Bergström, A., *et al.*, 'A Genomic History of Aboriginal Australia', *Nature*, 538, 7624 (2016), pp. 207–214.

Mallick, S., Li, H., Lipson, M., Mathieson, I., Gymrek, M., Racimo, F., Zhao, M., *et al.*, 'The Simons Genome Diversity Project: 300 Genomes from 142 Diverse Populations', *Nature*, 538, 7624 (2016), pp. 201–206.

McColl, H., Racimo, F., Vinner, L., Demeter, F., Gakuhari, T., Moreno-Mayar, J. V., Van Driem, G., *et al.*, 'The Prehistoric Peopling of Southeast Asia', *Science*, 361, 6397 (2018), pp. 88–92.

McDougall, I., Brown, F. H., Fleagle, J. G., 'Stratigraphic Placement and Age of Modern Humans from Kibish, Ethiopia', *Nature*, 433, 7027 (2005), pp. 733–736.

Mendes, M., Alvim, I., Borda, V., Tarazona-Santos, E., 'The History behind the Mosaic of the Americas', *Current Opinion in Genetics & Development*, 62 (2020), pp. 72–77.

Metspalu, M., Mondal, M., Chaubey, G., 'The Genetic Makings of South Asia', *Current Opinion in Genetics & Development*, 53 (2018), pp. 128–133.

Molinaro, L., Pagani, L., 'Human Evolutionary History of Eastern Africa', *Current Opinion in Genetics & Development*, 53 (2018), pp. 134–139.

Montano, V., Ferri, G., Marcari, V., Batini, C., Anyaele, O., Destro-Bisol, G., Comas, D., 'The Bantu Expansion Revisited: A New Analysis of Y Chromosome Variation in Central Western Africa', *Molecular Ecology*, 20, 13 (2011), pp. 2693–2708.

Moreno-Mayar, J. V., Vinner, L., de Barros Damgaard, P., de la Fuente, C., Chan, J., Spence, J. P., Allentoft, M. E., *et al.*, 'Early Human Dispersals within the Americas', *Science*, 362, eaav2621 (2018).

Nielsen, R., Akey, J. M., Jakobsson, M., Pritchard, J. K., Tishkoff, S., Willerslev, E., 'Tracing the Peopling of the World through Genomics', *Nature*, 541, 7637 (2017), pp. 302–310.

Novembre, J., Johnson, T., Bryc, K., Kutalik, Z., Boyko, A. R., Auton, A., Indap, A., *et al.*, 'Genes Mirror Geography within Europe', *Nature*, 456, 7218 (2008), pp. 98–101.

Novembre, J., Ramachandran, S., 'Perspectives on Human Population Structure at the Cusp of the Sequencing Era', *Annual Review of Genomics and Human Genetics*, 12 (2011), pp. 245–274.

Pagani, L., Lawson, D. J., Jagoda, E., Morseburg, A., Eriksson, A., Mitt, M., Clemente, F., *et al.*, 'Genomic Analyses Inform on Migration Events during the Peopling of Eurasia', *Nature*, 538, 7624 (2016), pp. 238–242.

Patin, E., Laval, G., Barreiro, L. B., Salas, A., Semino, O., Santachiara-Benerecetti, S., Kidd, K. K., *et al.*, 'Inferring the Demographic History of African Farmers and Pygmy Hunter-Gatherers Using a Multilocus Resequencing Data Set', *PLOS Genetics*, 5, 4 (2009), e1000448.

Patin, E., Lopez, M., Grollemund, R., Verdu, P., Harmant, C., Quach, H., Laval, G., *et al.*, 'Dispersals and Genetic Adaptation of Bantu-Speaking Populations in Africa and North America', *Science*, 356, 6337 (2017), pp. 543–546.

Patin, E., Quintana-Murci, L., 'The Demographic and Adaptive History of Central African Hunter-Gatherers and Farmers', *Current Opinion in Genetics & Development*, 53 (2018), pp. 90–97.

Patin, E., Siddle, K. J., Laval, G., Quach, H., Harmant, C., Becker, N., Froment, A., *et al.*, 'The Impact of Agricultural Emergence on the Genetic History of African Rainforest Hunter-Gatherers and Agriculturalists', *Nature Communications*, 5 (2014), p. 3163.

Patterson, N., Moorjani, P., Luo, Y., Mallick, S., Rohland, N., Zhan, Y., Genschoreck, T., Webster, T., Reich, D., 'Ancient Admixture in Human History', *Genetics*, 192, 3 (2012), pp. 1065–1093.

Petersen, D. C., Libiger, O., Tindall, E. A., Hardie, R. A., Hannick, L. I., Glashoff, R. H., Mukerji, M., *et al.*, 'Complex Patterns of Genomic Admixture within Southern Africa', *PLOS Genetics*, 9, 3 (2013), e1003309.

Phillipson, D. W., 'The Chronology of the Iron Age in Bantu Africa', *Journal of African History*, 16 (1975), pp. 321–342.

Phillipson, D. W., *African Archaeology*, Cambridge, Cambridge University Press, 1993.

Pickrell, J. K., Patterson, N., Barbieri, C., Berthold, F., Gerlach, L., Güldemann, T., Kure, B., *et al.*, 'The Genetic Prehistory of Southern Africa', *Nature Communications*, 3 (2012), pp. 1143.

Posth, C., Nagele, K., Colleran, H., Valentin, F., Bedford, S., Kami, K. W., Shing, R., *et al.*, 'Language Continuity despite Population Replacement in Remote Oceania', *Nature Ecology & Evolution*, 2, 4 (2018), pp. 731–740.

Posth, C., Nakatsuka, N., Lazaridis, I., Skoglund, P., Mallick, S., Lamnidis, T. C., Rohland, N., *et al.*, 'Reconstructing the Deep Population History of Central and South America', *Cell*, 175, 5 (2018), pp. 1185–1197, e22.

Pugach, I., Duggan, A. T., Merriwether, D. A., Friedlaender, F. R., Friedlaender, J. S., Stoneking, M., 'The Gateway from Near into Remote Oceania: New Insights from Genome-Wide Data', *Molecular Biology and Evolution*, 35, 4 (2018), pp. 871–886.

Qin, P., Stoneking, M., 'Denisovan Ancestry in East Eurasian and Native American Populations', *Molecular Biology and Evolution*, 32, 10 (2015), pp. 2665–2674.

Quintana-Murci, L., Harmant, C., Quach, H., Balanovsky, O., Zaporozhchenko, V., Bormans, C., van Helden, P. D., Hoal, E. G., Behar, D. M., 'Strong Maternal Khoisan Contribution to the South African Coloured Population: A Case of Gender-Biased Admixture', *American Journal of Human Genetics*, 86, 4 (2010), pp. 611–620.

Quintana-Murci, L., Quach, H., Harmant, C., Luca, F., Massonnet, B., Patin, E., Sica, L., *et al.*, 'Maternal Traces of Deep Common Ancestry and Asymmetric Gene Flow between Pygmy Hunter-Gatherers and Bantu-Speaking Farmers', *Proceedings of the National Academy of Sciences USA*, 105, 5 (2008), pp. 1596–1601.

Quintana-Murci, L., Semino, O., Bandelt, H.-J., Passarino, G., McElreavey, K., Santachiara-Benerecetti, A. S., 'Genetic Evidence of an Early Exit of *Homo Sapiens Sapiens* from Africa through Eastern Africa', *Nature Genetics*, 23, 4 (1999), pp. 437–441.

Raghavan, M., Skoglund, P., Graf, K. E., Metspalu, M., Albrechtsen, A., Moltke, I., Rasmussen, S., *et al.*, 'Upper Palaeolithic Siberian Genome Reveals Dual Ancestry of Native Americans', *Nature*, 505, 7481 (2014), pp. 87–91.

Rasmussen, M., Anzick, S. L., Waters, M. R., Skoglund, P., DeGiorgio, M., Stafford, T. W., Jr, Rasmussen, S., *et al.*, 'The Genome of a Late Pleistocene Human from a Clovis Burial Site in Western Montana', *Nature*, 506, 7487 (2014), pp. 225–229.

Reich, D., Patterson, N., Campbell, D., Tandon, A., Mazieres, S., Ray, N., Parra, M. V., *et al.*, 'Reconstructing Native American Population History', *Nature*, 488, 7411 (2012), pp. 370–374.

Reich, D., Patterson, N., Kircher, M., Delfin, F., Nandineni, M. R., Pugach, I., Ko, A. M., *et al.*, 'Denisova Admixture and the First Modern Human Dispersals into Southeast Asia and Oceania', *American Journal of Human Genetics*, 89, 4 (2011), pp. 516–528.

Richter, D., Grün, R., Joannes-Boyau, R., Steele, T. E., Amani, F., Rué, M., Fernandes, P., *et al.*, 'The Age of the Hominin Fossils from Jebel Irhoud,

Morocco, and the Origins of the Middle Stone Age', *Nature*, 546, 7657 (2017), pp. 293–296.

Saint Pierre, A., Giemza, J., Alves, I., Karakachoff, M., Gaudin, M., Amouyel, P., Dartigues, J. F., *et al.*, 'The Genetic History of France', *European Journal of Human Genetics*, 28, 7 (2020), pp. 853–865.

Sankararaman, S., Mallick, S., Dannemann, M., Prüfer, K., Kelso, J., Pääbo, S., Patterson, N., Reich, D., 'The Genomic Landscape of Neanderthal Ancestry in Present-Day Humans', *Nature*, 507, 7492 (2014), pp. 354–357.

Sankararaman, S., Mallick, S., Patterson, N., Reich, D., 'The Combined Landscape of Denisovan and Neanderthal Ancestry in Present-Day Humans', *Current Biology*, 26, 9 (2016), pp. 1241–1247.

Schlebusch, C. M., Jakobsson, M., 'Tales of Human Migration, Admixture, and Selection in Africa', *Annual Review of Genomics and Human Genetics*, 19 (2018), pp. 405–428.

Schlebusch, C. M., Skoglund, P., Sjödin, P., Gattepaille, L. M., Hernandez, D., Jay, F., Li, S., *et al.*, 'Genomic Variation in Seven Khoe-San Groups Reveals Adaptation and Complex African History', *Science*, 338, 6105 (2012), pp. 374–379.

Schuster, S. C., Miller, W., Ratan, A., Tomsho, L. P., Giardine, B., Kasson, L. R., Harris, R. S., *et al.*, 'Complete Khoisan and Bantu Genomes from Southern Africa', *Nature*, 463, 7283 (2010), pp. 943–947.

Serra-Vidal, G., Lucas-Sanchez, M., Fadhlaoui-Zid, K., Bekada, A., Zalloua, P., Comas, D., 'Heterogeneity in Palaeolithic Population Continuity and Neolithic Expansion in North Africa', *Current Biology*, 29, 22 (2019), pp. 3953–3959, e4.

Sirugo, G., Williams, S. M., Tishkoff, S. A., 'The Missing Diversity in Human Genetic Studies', *Cell*, 177, 4 (2019), p. 1080.

Skoglund, P., Jakobsson, M., 'Archaic Human Ancestry in East Asia', *Proceedings of the National Academy of Sciences USA*, 108, 45 (2011), pp. 18301–18306.

Skoglund, P., Mallick, S., Bortolini, M. C., Chennagiri, N., Hunemeier, T., Petzl-Erler, M. L., Salzano, F. M., Patterson, N., Reich, D., 'Genetic Evidence for Two Founding Populations of the Americas', *Nature*, 525, 7567 (2015), pp. 104–108.

Skoglund, P., Malmström, H., Raghavan, M., Storå, J., Hall, P., Willerslev, E., Gilbert, M. T. P., Götherström, A., Jakobsson, M., 'Origins and Genetic Legacy of Neolithic Farmers and Hunter-Gatherers in Europe', *Science*, 336, 6080 (2012), pp. 466–469.

Skoglund, P., Mathieson, I., 'Ancient Genomics of Modern Humans: The First Decade', *Annual Review of Genomics and Human Genetics*, 19 (2018), pp. 381–404.

Skoglund, P., Posth, C., Sirak, K., Spriggs, M., Valentin, F., Bedford, S., Clark, G. R., *et al.*, 'Genomic Insights into the Peopling of the Southwest Pacific', *Nature*, 538, 7626 (2016), pp. 510–513.

Skoglund, P., Reich, D., 'A Genomic View of the Peopling of the Americas', *Current Opinion in Genetics & Development*, 41 (2016), pp. 27–35.

Skoglund, P., Thompson, J. C., Prendergast, M. E., Mittnik, A., Sirak, K., Hajdinjak, M., Salie, T., *et al.*, 'Reconstructing Prehistoric African Population Structure', *Cell*, 171, 1 (S2017), pp. 59–71, e21.

Stringer, C. B., Andrews, P., 'Genetic and Fossil Evidence for the Origin of Modern Humans', *Science*, 239, 4845 (1988), pp. 1263–1268.

Stringer, C. B., Grün, R., Schwarcz, H. P., Goldberg, P., 'ESR dates for the Hominid Burial Site of Es Skhul in Israel', *Nature*, 338, 6218 (1989), pp. 756–758.

Tishkoff, S. A., Reed, F. A., Friedlaender, F. R., Ehret, C., Ranciaro, A., Froment, A., Hirbo, J. B., *et al.*, 'The Genetic Structure and History of Africans and African Americans', *Science*, 324, 5930 (2009), pp. 1035–1044.

Tishkoff, S. A., Williams, S. M., 'Genetic Analysis of African Populations: Human Evolution and Complex Disease', *Nature Review Genetics*, 3, 8 (2002), pp. 611–621.

Underhill, P. A., Shen, P., Lin, A. A., Jin, L., Passarino, G., Yang, W. H., Kauffman, E., *et al.*, 'Y Chromosome Sequence Variation and the History of Human Populations', *Nature Genetics*, 26, 3 (2000), pp. 358–361.

Vansina, J., 'Western Bantu Expansion', *Journal of African History*, 25 (1984), pp. 129–145.

Veeramah, K. R., Hammer, M. F., 'The Impact of Whole-Genome Sequencing on the Reconstruction of Human Population History', *Nature Review Genetics*, 15, 3 (2014), pp. 149–162.

Veeramah, K. R., Novembre, J., 'Demographic Events and Evolutionary Forces Shaping European Genetic Diversity', *Cold Spring Harbor Perspectives in Biology*, 6, 9 (2014), a008516.

Veeramah, K. R., Wegmann, D., Woerner, A., Mendez, F. L., Watkins, J. C., Destro-Bisol, G., Soodyall, H., Louie, L., Hammer, M. F., 'An Early Divergence of KhoeSan Ancestors from Those of Other Modern Humans Is Supported by an ABC-Based Analysis of Autosomal Resequencing Data', *Molecular Biology and Evolution*, 29, 2 (2012), pp. 617–630.

Verdu, P., 'African Pygmies', *Current Biology*, 26, 1 (2016), R12–R14.

Verdu, P., Austerlitz, F., Estoup, A., Vitalis, R., Georges, M., Thery, S., Froment, A., *et al.*, 'Origins and Genetic Diversity of Pygmy Hunter-Gatherers from Western Central Africa', *Current Biology*, 19, 4 (2009), pp. 312–318.

Verdu, P., Becker, N. S., Froment, A., Georges, M., Grugni, V., Quintana-Murci, L., Hombert, J. M., *et al.*, 'Sociocultural Behavior, Sex-Biased Admixture, and Effective Population Sizes in Central African Pygmies and Non-Pygmies', *Molecular Biology and Evolution*, 30, 4 (2013), pp. 918–937.

Vernot, B., Tucci, S., Kelso, J., Schraiber, J. G., Wolf, A. B., Gittelman, R. M., Dannemann, M., *et al.*, 'Excavating Neandertal and Denisovan DNA from the Genomes of Melanesian Individuals', *Science*, 352, 6282 (2016), pp. 235–239.

Wang, C. C., Yeh, H. Y., Popov, A. N., Zhang, H. Q., Matsumura, H., Sirak, K., Cheronet, O., *et al.*, 'Genomic Insights into the Formation of Human Populations in East Asia', *Nature*, 591, 7850 (2021), pp. 413–419.

Wolpoff, M. H., Wu, X. Z., Thorne, A. G., 'Modern *Homo Sapiens* Origins: A General Theory of Hominid Evolution Involving the Fossil Evidence from East Asia', in F. H. Smith and F. Spencer (eds.), *The Origins of Modern Humans: A World Survey of the Fossil Evidence*, New York, Alan Liss, 1984, pp. 411–483.

Yang, M. A., Fan, X., Sun, B., Chen, C., Lang, J., Ko, Y.-C., Tsang, C.-H., *et al.*, 'Ancient DNA Indicates Human Population Shifts and Admixture in Northern and Southern China', *Science*, 369, 6501 (2020), pp. 282–288.

Yang, M. A., Gao, X., Theunert, C., Tong, H., Aximu-Petri, A., Nickel, B., Slatkin, M., *et al.*, '40,000-Year-Old Individual from Asia Provides Insight into Early Population Structure in Eurasia', *Current Biology*, 27, 20 (2017), pp. 3202–3208, e9.

3. Adaptation and Environment

Alkorta-Aranburu, G., Beall, C. M., Witonsky, D. B., Gebremedhin, A., Pritchard, J. K., Di Rienzo, A., 'The Genetic Architecture of Adaptations to High Altitude in Ethiopia', *PLOS Genetics*, 8, 12 (2012), e1003110.

Allison, A. C., 'Protection Afforded by Sickle-Cell Trait against Subtertian Malarial Infection', *British Medical Journal*, 1, 4857 (1954), pp. 290–294.

Andrés, A. M. 'Balancing Selection in the Human Genome', *in Encyclopedia of Life Sciences*, Chichester, John Wiley and Sons, 2011.

Asthana, S., Schmidt, S., Sunyaev, S., 'A Limited Role for Balancing Selection', *Trends in Genetics*, 21, 1 (2005), pp. 30–32.

Beall, C. M., 'Detecting Natural Selection in High-Altitude Human Populations', *Respiratory Physiology & Neurobiology*, 158, 2–3 (2007), pp. 161–171.

Becker, N. S., Verdu, P., Froment, A., Le Bomin, S., Pagezy, H., Bahuchet, S., Heyer, E., 'Indirect Evidence for the Genetic Determination of Short Stature in African Pygmies', *American Journal of Physical Anthropology*, 145, 3 (2011), pp. 390–401.

Becker, N. S., Verdu, P., Georges, M., Duquesnoy, P., Froment, A., Amselem, S., Le Bouc, Y., Heyer, E., 'The Role of *GHR* and *IGF1* Genes in the Genetic Determination of African Pygmies' Short Stature', *European Journal of Human Genetics*, 21, 6 (2013), pp. 653–658.

Berg, J. J., Coop, G., 'A Population Genetic Signal of Polygenic Adaptation', *PLOS Genetics*, 10, 8 (2014), e1004412.

Berg, J. J., Harpak, A., Sinnott-Armstrong, N., Joergensen, A. M., Mostafavi, H., Field, Y., Boyle, E. A., *et al.*, 'Reduced Signal for Polygenic Adaptation of Height in UK Biobank', *eLife*, 8 (2019), e39725.

Bergey, C. M., Lopez, M., Harrison, G. F., Patin, E., Cohen, J. A., Quintana-Murci, L., Barreiro, L. B., Perry, G. H., 'Polygenic Adaptation and Convergent Evolution on Growth and Cardiac Genetic Pathways in African and Asian Rainforest Hunter-Gatherers', *Proceedings of the National Academy of Sciences USA*, 115, 48 (2018), E11256–E11263.

Bersaglieri, T., Sabeti, P. C., Patterson, N., Vanderploeg, T., Schaffner, S. F., Drake, J. A., Rhodes, M., Reich, D. E., Hirschhorn, J. N., 'Genetic Signatures of Strong Recent Positive Selection at the Lactase Gene', *American Journal of Human Genetics*, 74, 6 (2004), pp. 1111–1120.

Bigham, A., Bauchet, M., Pinto, D., Mao, X., Akey, J. M., Mei, R., Scherer, S. W., *et al.*, 'Identifying Signatures of Natural Selection in Tibetan and Andean Populations Using Dense Genome Scan Data', *PLOS Genetics*, 6, 9 (2010), e1001116.

Bitarello, B. D., de Filippo, C., Teixeira, J. C., Schmidt, J. M., Kleinert, P., Meyer, D., Andrés, A. M., 'Signatures of Long-Term Balancing Selection in Human Genomes', *Genome Biology and Evolution*, 10, 3 (2018), pp. 939–955.

Bustamante, C. D., Fledel-Alon, A., Williamson, S., Nielsen, R., Hubisz, M. T., Glanowski, S., Tanenbaum, D. M., *et al.*, 'Natural Selection on Protein-Coding Genes in the Human Genome', *Nature*, 437, 7062 (2005), pp. 1153–1157.

Bycroft, C., Freeman, C., Petkova, D., Band, G., Elliott, L. T., Sharp, K., Motyer, A., *et al.*, 'The UK Biobank Resource with Deep Phenotyping and Genomic Data', *Nature*, 562, 7726 (2018), pp. 203–209.

Charlesworth, D., 'Balancing Selection and Its Effects on Sequences in Nearby Genome Regions', *PLOS Genetics*, 2, 4 (2006), e64.

Clemente, F. J., Cardona, A., Inchley, C. E., Peter, B. M., Jacobs, G., Pagani, L., Lawson, D. J., *et al.*, 'A Selective Sweep on a Deleterious Mutation in *CPT1A* in Arctic Populations', *American Journal of Human Genetics*, 95, 5 (2014), pp. 584–589.

Crawford, N. G., Kelly, D. E., Hansen, M. E. B., Beltrame, M. H., Fan, S., Bowman, S. L., Jewett, E., *et al.*, 'Loci Associated with Skin Pigmentation Identified in African Populations', *Science*, 358, eaan8433 (2017).

de Filippo, C., Key, F. M., Ghirotto, S., Benazzo, A., Meneu, J. R., Weihmann, A., NISC Comparative Sequence Program, *et al.*, 'Recent Selection Changes in Human Genes under Long-Term Balancing Selection', *Molecular Biology and Evolution*, 33, 6 (2016), pp. 1435–1447.

DeGiorgio, M., Lohmueller, K. E., Nielsen, R., 'A Model-Based Approach for Identifying Signatures of Ancient Balancing Selection in Genetic Data', *PLOS Genetics*, 10, 8 (2014), e1004561.

Enard, W., Przeworski, M., Fisher, S. E., Lai, C. S. L., Wiebe, V., Kitano, T., Monaco, A. P., Pääbo, S., 'Molecular Evolution of *FOXP2*, a Gene Involved in Speech and Language', *Nature*, 418, 6900 (2002), pp. 869–872.

Enattah, N. S., Jensen, T. G., Nielsen, M., Lewinski, R., Kuokkanen, M., Rasinpera, H., El-Shanti, H., *et al.*, 'Independent Introduction of Two Lactase-Persistence Alleles into Human Populations Reflects Different History of Adaptation to Milk Culture', *American Journal of Human Genetics*, 82, 1 (2008), pp. 57–72.

Enattah, N. S., Sahi, T., Savilahti, E., Terwilliger, J. D., Peltonen, L., Järvelä, I., 'Identification of a Variant Associated with Adult-Type Hypolactasia', *Nature Genetics*, 30, 2 (2002), pp. 233–237.

Fan, S., Hansen, M. E., Lo, Y., Tishkoff, S. A., 'Going Global by Adapting Local: A Review of Recent Human Adaptation', *Science*, 354, 6308 (2016), pp. 54–59.

Ferrer-Admetlla, A., Liang, M., Korneliussen, T., Nielsen, R., 'On Detecting Incomplete Soft or Hard Selective Sweeps Using Haplotype Structure', *Molecular Biology and Evolution*, 31, 5 (2014), pp. 1275–1291.

Field, Y., Boyle, E. A., Telis, N., Gao, Z., Gaulton, K. J., Golan, D., Yengo, L., *et al.*, 'Detection of Human Adaptation during the Past 2000 Years', *Science*, 354, 6313 (2016), pp. 760–764.

Fisher, R. A., Ford, E. B., Huxley, J., 'Taste-Testing the Anthropoid Apes', *Nature*, 144 (1939), p. 750.

Fumagalli, M., Moltke, I., Grarup, N., Racimo, F., Bjerregaard, P., Jorgensen, M. E., Korneliussen, T. S., *et al.*, 'Greenlandic Inuit Show Genetic Signatures of Diet and Climate Adaptation', *Science*, 349, 6254 (2015), pp. 1343–1347.

Han, J., Kraft, P., Nan, H., Guo, Q., Chen, C., Qureshi, A., Hankinson, S. E., *et al.*, 'A Genome-Wide Association Study Identifies Novel Alleles Associated with Hair Color and Skin Pigmentation', *PLOS Genetics*, 4, 5 (2008), e1000074.

Hancock, A. M., Witonsky, D. B., Gordon, A. S., Eshel, G., Pritchard, J. K., Coop, G., Di Rienzo, A., 'Adaptations to Climate in Candidate Genes for Common Metabolic Disorders', *PLOS Genetics*, 4, 2 (2008), e32.

Harris, E. E., Meyer, D., 'The Molecular Signature of Selection Underlying Human Adaptations', *American Journal of Physical Anthropology*, Suppl. 43 (2006), pp. 89–130.

Hernandez, R. D., Kelley, J. L., Elyashiv, E., Melton, S. C., Auton, A., McVean, G., Sella, G., Przeworski, M., 'Classic Selective Sweeps Were

Rare in Recent Human Evolution', *Science*, 331, 6019 (2011), pp. 920–924.

Hsieh, P., Veeramah, K. R., Lachance, J., Tishkoff, S. A., Wall, J. D., Hammer, M. F., Gutenkunst, R. N., 'Whole-Genome Sequence Analyses of Western Central African Pygmy Hunter-Gatherers Reveal a Complex Demographic History and Identify Candidate Genes under Positive Natural Selection', *Genome Research*, 26, 3 (2016), pp. 279–290.

Huerta-Sánchez, E., Jin, X., Asan Bianba, Z., Peter, B. M., Vinckenbosch, N., Liang, Y., *et al.*, 'Altitude Adaptation in Tibetans Caused by Introgression of Denisovan-Like DNA', *Nature*, 512, 7513 (2014), pp. 194–197.

Ilardo, M. A., Moltke, I., Korneliussen, T. S., Cheng, J., Stern, A. J., Racimo, F., de Barros Damgaard, P., *et al.*, 'Physiological and Genetic Adaptations to Diving in Sea Nomads', *Cell*, 173, 3 (2018), pp. 569–80, e15.

Ilardo, M., Nielsen, R., 'Human Adaptation to Extreme Environmental Conditions', *Current Opinion in Genetics & Development*, 53 (2018), pp. 77–82.

Itan, Y., Jones, B. L., Ingram, C. J., Swallow, D. M., Thomas, M. G., 'A Worldwide Correlation of Lactase Persistence Phenotype and Genotypes', *BMC Ecology and Evolution*, 10 (2010), p. 36.

Itan, Y., Powell, A., Beaumont, M. A., Burger, J., Thomas, M. G., 'The Origins of Lactase Persistence in Europe', *PLOS Computational Biology*, 5, 8 (2009), e1000491.

Jablonski, N. G., Chaplin, G., 'The Evolution of Human Skin Coloration', *Journal of Human Evolution*, 39, 1 (2000), pp. 57–106.

Jablonski, N. G., Chaplin, G., 'Human Skin Pigmentation, Migration and Disease Susceptibility', *Philosophical Transactions of the Royal Society of London B: Biological Sciences*, 367, 1590 (2012), pp. 785–792.

Jablonski, N. G., Chaplin, G., 'The Colours of Humanity: The Evolution of Pigmentation in the Human Lineage', *Philosophical Transactions of the Royal Society of London B: Biological Sciences*, 372, 20160349 (2017).

Jarvis, J. P., Scheinfeldt, L. B., Soi, S., Lambert, C., Omberg, L., Ferwerda, B., Froment, A., *et al.*, 'Patterns of Ancestry, Signatures of Natural Selection, and Genetic Association with Stature in Western African Pygmies', *PLOS Genetics*, 8, 4 (2012), e1002641.

Jeong, C., Alkorta-Aranburu, G., Basnyat, B., Neupane, M., Witonsky, D. B., Pritchard, J. K., Beall, C. M., Di Rienzo, A., 'Admixture Facilitates Genetic Adaptations to High Altitude in Tibet', *Nature Communications*, 5 (2014), p. 3281.

Jeong, C., Di Rienzo, A., 'Adaptations to Local Environments in Modern Human Populations', *Current Opinion in Genetics & Development*, 29 (2014), pp. 1–8.

Key, F. M., Abdul-Aziz, M. A., Mundry, R., Peter, B. M., Sekar, A., D'Amato, M., Dennis, M. Y., Schmidt, J. M., Andrés, A. M., 'Human Local Adaptation of the TRPM8 Cold Receptor along a Latitudinal Cline', *PLOS Genetics*, 14, 5 (2018), e1007298.

Key, F. M., Teixeira, J. C., de Filippo, C., Andrés, A. M., 'Advantageous Diversity Maintained by Balancing Selection in Humans', *Current Opinion in Genetics & Development*, 29 (2014), pp. 45–51.

Lachance, J., Tishkoff, S. A., 'Population Genomics of Human Adaptation', *Annual Review of Ecology, Evolution and Systematics*, 44 (2013), pp. 123–143.

Lachance, J., Vernot, B., Elbers, C. C., Ferwerda, B., Froment, A., Bodo, J. M., Lema, G., *et al.*, 'Evolutionary History and Adaptation from High-Coverage Whole-Genome Sequences of Diverse African Hunter-Gatherers', *Cell*, 150, 3 (2012), pp. 457–469.

Lango Allen, H., Estrada, K., Lettre, G., Berndt, S. I., Weedon, M. N., Rivadeneira, F., Willer, C. J., *et al.*, 'Hundreds of Variants Clustered in Genomic Loci and Biological Pathways Affect Human Height', *Nature*, 467, 7317 (2010), pp. 832–838.

Leffler, E. M., Gao, Z., Pfeifer, S., Ségurel, L., Auton, A., Venn, O., Bowden, R., *et al.*, 'Multiple Instances of Ancient Balancing Selection Shared between Humans and Chimpanzees', *Science*, 339, 6127 (2013), pp. 1578–1582.

Lohmueller, K. E., 'The Distribution of Deleterious Genetic Variation in Human Populations', *Current Opinion in Genetics & Development*, 29 (2014), pp. 139–146.

Lohmueller, K. E., Indap, A. R., Schmidt, S., Boyko, A. R., Hernandez, R. D., Hubisz, M. J., Sninsky, J. J., *et al.*, 'Proportionally More Deleterious Genetic Variation in European than in African Populations', *Nature*, 451, 7181 (2008), pp. 994–997.

Lopez, M., Choin, J., Sikora, M., Siddle, K., Harmant, C., Costa, H. A., Silvert, M., *et al.*, 'Genomic Evidence for Local Adaptation of Hunter-Gatherers to the African Rainforest', *Current Biology*, 29, 17 (2019), pp. 2926–2935, e4.

Lopez, M., Kousathanas, A., Quach, H., Harmant, C., Mouguiama-Daouda, P., Hombert, J. M., Froment, A., *et al.*, 'The Demographic History and Mutational Load of African Hunter-Gatherers and Farmers', *Nature Ecology & Evolution*, 2, 4 (2018), pp. 721–730.

Martin, A. R., Lin, M., Granka, J. M., Myrick, J. W., Liu, X., Sockell, A., Atkinson, E. G., *et al.*, 'An Unexpectedly Complex Architecture for Skin Pigmentation in Africans', *Cell*, 171, 6 (2017), pp. 1340–1353, e14.

McEvoy, B., Beleza, S., Shriver, M. D., 'The Genetic Architecture of Normal Variation in Human Pigmentation: An Evolutionary Perspective and Model', *Human Molecular Genetics*, 15, 2 (2006), R176–R181.

Mendizabal, I., Marigorta, U. M., Lao, O., Comas, D., 'Adaptive Evolution of Loci Covarying with the Human African Pygmy Phenotype', *Human Genetics*, 131, 8 (2012), pp. 1305–1317.

Myles, S., Somel, M., Tang, K., Kelso, J., Stoneking, M., 'Identifying Genes Underlying Skin Pigmentation Differences among Human Populations', *Human Genetics*, 120, 5 (2007), pp. 613–621.

Nielsen, R., 'Molecular Signatures of Natural Selection', *Annual Review of Genetics*, 39 (2005), pp. 197–218.

Nielsen, R., Hellmann, I., Hubisz, M., Bustamante, C., Clark, A. G., 'Recent and Ongoing Selection in the Human Genome', *Nature Review Genetics*, 8, 11 (2007), pp. 857–868.

Nielsen, R., Williamson, S., Kim, Y., Hubisz, M. J., Clark, A. G., Bustamante, C., 'Genomic Scans for Selective Sweeps Using SNP Data', *Genome Research*, 15, 11 (2005), pp. 1566–1575.

Norton, H. L., Kittles, R. A., Parra, E., McKeigue, P., Mao, X., Cheng, K., Canfield, V. A., *et al.*, 'Genetic Evidence for the Convergent Evolution of Light Skin in Europeans and East Asians', *Molecular Biology and Evolution*, 24, 3 (2007), pp. 710–722.

Olalde, I., Allentoft, M. E., Sanchez-Quinto, F., Santpere, G., Chiang, C. W., DeGiorgio, M., Prado-Martinez, J., *et al.*, 'Derived Immune and

Ancestral Pigmentation Alleles in a 7,000-Year-Old Mesolithic European', *Nature*, 507, 7491 (2014), pp. 225–228.

Patin, E., Quintana-Murci, L., 'Demeter's Legacy: Rapid Changes to Our Genome Imposed by Diet', *Trends in Ecology & Evolution*, 23, 2 (2008), pp. 56–59.

Pavan, W. J., Sturm, R. A., 'The Genetics of Human Skin and Hair Pigmentation', *Annual Review of Genomics and Human Genetics*, 20 (2019), pp. 41–72.

Pemberton, T. J., Verdu, P., Becker, N. S., Willer, C. J., Hewlett, B. S., Le Bomin, S., Froment, A., Rosenberg, N. A., Heyer, E., 'A Genome Scan for Genes Underlying Adult Body Size Differences between Central African Hunter-Gatherers and Farmers', *Human Genetics*, 137, 6–7 (2018), pp. 487–509.

Peng, Y., Yang, Z., Zhang, H., Cui, C., Qi, X., Luo, X., Tao, X., *et al.*, 'Genetic Variations in Tibetan Populations and High-Altitude Adaptation at the Himalayas', *Molecular Biology and Evolution*, 28, 2 (2011), pp. 1075–1081.

Perry, G. H., Dominy, N. J., 'Evolution of the Human Pygmy Phenotype', *Trends in Ecology & Evolution*, 24, 4 (2009), pp. 218–225.

Perry, G. H., Foll, M., Grenier, J.-C., Patin, E., Nédélec, Y., Pacis, A., Barakatt, M., *et al.*, 'Adaptive, Convergent Origins of the Pygmy Phenotype in African Rainforest Hunter-Gatherers', *Proceedings of the National Academy of Sciences USA*, 111, 35 (2014), E3596–E3603.

Peter, B. M., Huerta-Sánchez, E., Nielsen, R., 'Distinguishing between Selective Sweeps from Standing Variation and from a *De Novo* Mutation', *PLOS Genetics*, 8, 10 (2012), e1003011.

Pritchard, J. K., Di Rienzo, A., 'Adaptation – Not by Sweeps Alone', *Nature Review Genetics*, 11, 10 (2010), pp. 665–667.

Pritchard, J. K., Pickrell, J. K., Coop, G., 'The Genetics of Human Adaptation: Hard Sweeps, Soft Sweeps, and Polygenic Adaptation', *Current Biology*, 20, 4 (2010), R208–R215.

Przeworski, M., Coop, G., Wall, J. D., 'The Signature of Positive Selection on Standing Genetic Variation', *Evolution*, 59, 11 (2005), pp. 2312–2323.

Quach, H., Quintana-Murci, L., 'Living in an Adaptive World: Genomic Dissection of the Genus *Homo* and Its Immune Response', *Journal of Experimental Medicine*, 214, 4 (2017), pp. 877–894.

Quintana-Murci, L., 'Genetic and Epigenetic Variation of Human Populations: An Adaptive Tale', *Comptes Rendues Biologies*, 339, 7–8 (2016), pp. 278–283.

Ranciaro, A., Campbell, M. C., Hirbo, J. B., Ko, W. Y., Froment, A., Anagnostou, P., Kotze, M. J., *et al.*, 'Genetic Origins of Lactase Persistence and the Spread of Pastoralism in Africa', *American Journal of Human Genetics*, 94, 4 (2014), pp. 496–510.

Robinson, M. R., Hemani, G., Medina-Gomez, C., Mezzavilla, M., Esko, T., Shakhbazov, K., Powell, J. E., *et al.*, 'Population Genetic Differentiation of Height and Body Mass Index across Europe', *Nature Genetics*, 47, 11 (2015), pp. 1357–1362.

Sabeti, P. C., Reich, D. E., Higgins, J. M., Levine, H. Z., Richter, D. J., Schaffner, S. F., Gabriel, S. B., *et al.*, 'Detecting Recent Positive Selection in the Human Genome from Haplotype Structure', *Nature*, 419, 6909 (2002), pp. 832–837.

Sabeti, P. C., Schaffner, S. F., Fry, B., Lohmueller, J., Varilly, P., Shamovsky, O., Palma, A., *et al.*, 'Positive Natural Selection in the Human Lineage', *Science*, 312, 5780 (2006), pp. 1614–1620.

Sabeti, P. C., Varilly, P., Fry, B., Lohmueller, J., Hostetter, E., Cotsapas, C., Xie, X., *et al.*, 'Genome-Wide Detection and Characterization of Positive Selection in Human Populations', *Nature*, 449, 7164 (2007), pp. 913–918.

Scheinfeldt, L. B., Soi, S., Thompson, S., Ranciaro, A., Woldemeskel, D., Beggs, W., Lambert, C., *et al.*, 'Genetic Adaptation to High Altitude in the Ethiopian Highlands', *Genome Biology*, 13, 1 (2012), R1.

Scheinfeldt, L. B., Tishkoff, S. A., 'Recent Human Adaptation: Genomic Approaches, Interpretation and Insights', *Nature Review Genetics*, 14, 10 (2013), pp. 692–702.

Ségurel, L., Bon, C., 'On the Evolution of Lactase Persistence in Humans', *Annual Review of Genomics and Human Genetics*, 18 (2017), pp. 297–319.

Ségurel, L., Gao, Z., Przeworski, M., 'Ancestry Runs Deeper than Blood: The Evolutionary History of ABO Points to Cryptic Variation of Functional Importance', *Bioessays*, 35, 10 (2013), pp. 862–867.

Ségurel, L., Thompson, E. E., Flutre, T., Lovstad, J., Venkat, A., Margulis, S. W., Moyse, J., *et al.*, 'The ABO Blood Group Is a Trans-Species Polymorphism in Primates', *Proceedings of the National Academy of Sciences USA*, 109, 45 (2012), pp. 18493–18498.

Simons, Y. B., Sella, G., 'The Impact of Recent Population History on the Deleterious Mutation Load in Humans and Close Evolutionary Relatives', *Current Opinion in Genetics & Development*, 41 (2016), pp. 150–158.

Simons, Y. B., Turchin, M. C., Pritchard, J. K., Sella, G., 'The Deleterious Mutation Load Is Insensitive to Recent Population History', *Nature Genetics*, 46, 3 (2014), pp. 220–224.

Simonson, T. S., Yang, Y., Huff, C. D., Yun, H., Qin, G., Witherspoon, D. J., Bai, Z., *et al.*, 'Genetic Evidence for High-Altitude Adaptation in Tibet', *Science*, 329, 5987 (2010), pp. 72–75.

Sohail, M., Maier, R. M., Ganna, A., Bloemendal, A., Martin, A. R., Turchin, M. C., Chiang, C. W., *et al.*, 'Polygenic Adaptation on Height Is Overestimated Due to Uncorrected Stratification in Genome-Wide Association Studies', *eLife*, 8 (2019), e39702.

Sulem, P., Gudbjartsson, D. F., Stacey, S. N., Helgason, A., Rafnar, T., Magnusson, K. P., Manolescu, A., *et al.*, 'Genetic Determinants of Hair, Eye and Skin Pigmentation in Europeans', *Nature Genetics*, 39, 12 (2007), pp. 1443–1452.

Swallow, D. M., 'Genetics of Lactase Persistence and Lactose Intolerance', *Annual Review of Genetics*, 37 (2003), pp. 197–219.

Takahata, N., Satta, Y., Klein, J., 'Polymorphism and Balancing Selection at Major Histocompatibility Complex Loci', *Genetics*, 130, 4 (1992), pp. 925–938.

Tishkoff, S. A., Reed, F. A., Ranciaro, A., Voight, B. F., Babbitt, C. C., Silverman, J. S., Powell, K., *et al.*, 'Convergent Adaptation of Human Lactase Persistence in Africa and Europe', *Nature Genetics*, 39, 1 (2007), pp. 31–40.

Tucci, S., Vohr, S. H., McCoy, R. C., Vernot, B., Robinson, M. R., Barbieri, C., Nelson, B. J., *et al.*, 'Evolutionary History and Adaptation of a Human Pygmy Population of Flores Island, Indonesia', *Science*, 361, 6401 (2018), pp. 511–516.

Turchin, M. C., Chiang, C. W., Palmer, C. D., Sankararaman, S., Reich, D., Hirschhorn, J. N., 'Evidence of Widespread Selection on Standing Variation in Europe at Height-Associated SNPs', *Nature Genetics*, 44, 9 (2012), pp. 1015–1019.

Vitti, J. J., Grossman, S. R., Sabeti, P. C., 'Detecting Natural Selection in Genomic Data', *Annual Review of Genetics*, 47 (2013), pp. 97–120.

Wooding, S., Bufe, B., Grassi, C., Howard, M. T., Stone, A. C., Vazquez, M., Dunn, D. M., *et al.*, 'Independent Evolution of Bitter-Taste Sensitivity in Humans and Chimpanzees', *Nature*, 440, 7086 (2006), pp. 930–934.

Wooding, S., Kim, U.-K., Bamshad, M. J., Larsen, J., Jorde, L. B., Drayna, D., 'Natural Selection and Molecular Evolution in *PTC*, a Bitter-Taste Receptor Gene', *American Journal of Human Genetics*, 74, 4 (2004), pp. 637–646.

Yengo, L., Sidorenko, J., Kemper, K. E., Zheng, Z., Wood, A. R., Weedon, M. N., Frayling, T. M., *et al.*, 'Meta-Analysis of Genome-Wide Association Studies for Height and Body Mass Index in Approximately 700,000 Individuals of European Ancestry', *Human Molecular Genetics*, 27, 20 (2018), pp. 3641–3649.

Yi, X., Liang, Y., Huerta-Sánchez, E., Jin, X., Cuo, Z. X. P., Pool, J. E., Xu, X., *et al.*, 'Sequencing of 50 Human Exomes Reveals Adaptation to High Altitude', *Science*, 329, 5987 (2010), pp. 75–78.

4. Humans and Microbes

Abecasis, G. R., Altshuler, D., Auton, A., Brooks, L. D., Durbin, R. M., Gibbs, R. A., Hurles, M. E., McVean, G. A., 'A Map of Human Genome Variation from Population-Scale Sequencing', *Nature*, 467, 7319 (2010), pp. 1061–1073.

Abecasis, G. R., Auton, A., Brooks, L. D., DePristo, M. A., Durbin, R. M., Handsaker, R. E., Kang, H. M., Marth, G. T., McVean, G. A., 'An Integrated Map of Genetic Variation from 1,092 Human Genomes', *Nature*, 491, 7422 (2012), pp. 56–65.

Abel, L., Alcaïs, A., Schurr, E., 'The Dissection of Complex Susceptibility to Infectious Disease: Bacterial, Viral and Parasitic Infections', *Current Opinion in Immunology*, 30 (2014), pp. 72–78.

Agarwal, A., Guindo, A., Cissoko, Y., Taylor, J. G., Coulibaly, D., Koné, A., Kayentao, K., *et al.*, 'Hemoglobin C Associated with Protection from Severe Malaria in the Dogon of Mali, a West African Population with a Low Prevalence of Hemoglobin S', *Blood*, 96, 7 (2000), pp. 2358–2363.

Aitman, T. J., Cooper, L. D., Norsworthy, P. J., Wahid, F. N., Gray, J. K., Curtis, B. R., McKeigue, P. M., *et al.*, 'Malaria Susceptibility and *CD36* Mutation', *Nature*, 405, 6790 (2000), pp. 1015–1016.

Alcaïs, A., Abel, L., Casanova, J. L., 'Human Genetics of Infectious Diseases: Between Proof of Principle and Paradigm', *Journal of Clinical Investigation*, 119, 9 (2009), pp. 2506–2514.

Alcaïs, A., Quintana-Murci, L., Thaler, D. S., Schurr, E., Abel, L., Casanova, J. L., 'Life-Threatening Infectious Diseases of Childhood: Single-Gene Inborn Errors of Immunity?', *Annals of the New York Academy of Sciences*, 1214 (2010), pp. 18–33.

Allison, A. C., 'Protection Afforded by Sickle-Cell Trait against Subtertian Malarial Infection', *British Medical Journal*, 1, 4857 (1954), pp. 290–294.

Arenzana-Seisdedos, F., Parmentier, M., 'Genetics of Resistance to HIV Infection: Role of Co-Receptors and Co-Receptor Ligands', *Seminars in Immunology*, 18, 6 (2006), pp. 387–403.

Auton, A., Brooks, L. D., Durbin, R. M., Garrison, E. P., Kang, H. M., Korbel, J. O., Marchini, J. L., *et al.*, 'A Global Reference for Human Genetic Variation', *Nature*, 526, 7571 (2015), pp. 68–74.

Bakewell, M. A., Shi, P., Zhang, J., 'More Genes Underwent Positive Selection in Chimpanzee Evolution than in Human Evolution', *Proceedings of the National Academy of Sciences USA*, 104, 18 (2007), pp. 7489–7494.

Bamshad, M. J., Mummidi, S., Gonzalez, E., Ahuja, S. S., Dunn, D. M., Watkins, W. S., Wooding, S., *et al.*, 'A Strong Signature of Balancing Selection in the 5′ *Cis*-Regulatory Region of *CCR5*', *Proceedings of the National Academy of Sciences USA*, 99, 16 (2002), pp. 10539–10544.

Barreiro, L. B., Ben-Ali, M., Quach, H., Laval, G., Patin, E., Pickrell, J. K., Bouchier, C., *et al.*, 'Evolutionary Dynamics of Human Toll-Like Receptors and Their Different Contributions to Host Defense', *PLOS Genetics*, 5, 7 (2009), e1000562.

Barreiro, L. B., Patin, E., Neyrolles, O., Cann, H. M., Gicquel, B., Quintana-Murci, L., 'The Heritage of Pathogen Pressures and Ancient Demography in the Human Innate-Immunity *CD209/CD209L* Region', *American Journal of Human Genetics*, 77, 5 (2005), pp. 869–886.

Barreiro, L. B., Quintana-Murci, L., 'From Evolutionary Genetics to Human Immunology: How Selection Shapes Host Defence Genes', *Nature Review Genetics*, 11, 1 (2010), pp. 17–30.

Blekhman, R., Man, O., Herrmann, L., Boyko, A. R., Indap, A., Kosiol, C., Bustamante, C. D., Teshima, K. M., Przeworski, M., 'Natural Selection on Genes that Underlie Human Disease Susceptibility', *Current Biology*, 18, 12 (2008), pp. 883–889.

Boisson-Dupuis, S., Ramirez-Alejo, N., Li, Z., Patin, E., Rao, G., Kerner, G., Lim, C. K., *et al.*, 'Tuberculosis and Impaired IL-23-Dependent IFN-γ Immunity in Humans Homozygous for a Common *TYK2* Missense Variant', *Science Immunology*, 3, eaau8714 (2018).

Brinkworth, J. F., Barreiro, L. B., 'The Contribution of Natural Selection to Present-Day Susceptibility to Chronic Inflammatory and Auto-immune Disease', *Current Opinion in Immunology*, 31 (2014), pp. 66–78.

Bustamante, C. D., Fledel-Alon, A., Williamson, S., Nielsen, R., Hubisz, M. T., Glanowski, S., Tanenbaum, D. M., *et al.*, 'Natural Selection on Protein-Coding Genes in the Human Genome', *Nature*, 437, 7062 (2005), pp. 1153–1157.

Cairns, J., *Matters of Life and Death*, Princeton, Princeton University Press, 1997.

Carrington, M., Martin, M. P., van Bergen, J., '*KIR-HLA* Intercourse in HIV Disease', *Trends in Microbiology*, 16, 12 (2008), pp. 620–627.

Casanova, J. L., 'Human Genetic Basis of Interindividual Variability in the Course of Infection', *Proceedings of the National Academy of Sciences USA*, 112, 51 (2015), E7118–E7127.

Casanova, J. L., Abel, L., 'The Human Model: A Genetic Dissection of Immunity to Infection in Natural Conditions', *Nature Reviews Immunology*, 4, 1 (2004), pp. 55–66.

Casanova, J. L., Abel, L., 'Inborn Errors of Immunity to Infection: The Rule Rather than the Exception', *Journal of Experimental Medicine*, 202, 2 (2005), pp. 197–201.

Casanova, J. L., Abel, L., 'Human Genetics of Infectious Diseases: A Unified Theory', *EMBO Journal*, 26, 4 (2007), pp. 915–922.

Casanova, J. L., Abel, L., 'Human Genetics of Infectious Diseases: Unique Insights into Immunological Redundancy', *Seminars in Immunology*, 36 (2018), pp. 1–12.

Casanova, J. L., Abel, L., Quintana-Murci, L., 'Human TLRs and IL-1Rs in Host Defense: Natural Insights from Evolutionary, Epidemiological, and Clinical Genetics', *Annual Review of Immunology*, 29 (2011), pp. 447–491.

Casanova, J. L., Abel, L., Quintana-Murci, L., 'Immunology Taught by Human Genetics', *Cold Spring Harbor Symposia on Quantitative Biology*, 78 (2013), pp. 157–172.

Chaix, R., Cao, C., Donnelly, P., 'Is Mate Choice in Humans MHC-Dependent?', *PLOS Genetics*, 4, 9 (2008), e1000184.

Chapman, S. J., Hill, A. V., 'Human Genetic Susceptibility to Infectious Disease', *Nature Review Genetics*, 13, 3 (2012), pp. 175–188.

Chimpanzee Sequencing and Analysis Consortium, 'Initial Sequence of the Chimpanzee Genome and Comparison with the Human Genome', *Nature*, 437, 7055 (2005), pp. 69–87.

Chitnis, C. E., Miller, L. H., 'Identification of the Erythrocyte Binding Domains of *Plasmodium Vivax* and *Plasmodium Knowlesi* Proteins Involved in Erythrocyte Invasion', *Journal of Experimental Medicine*, 180, 2 (1994), pp. 497–506.

Comas, I., Coscolla, M., Luo, T., Borrell, S., Holt, K. E., Kato-Maeda, M., Parkhill, J., *et al.*, 'Out-of-Africa Migration and Neolithic Coexpansion of *Mycobacterium Tuberculosis* with Modern Humans', *Nature Genetics*, 45, 10 (2013), pp. 1176–1182.

Cooke, G. S., Hill, A. V., 'Genetics of Susceptibility to Human Infectious Disease', *Nature Review Genetics*, 2, 12 (2001), pp. 967–977.

Dandine-Roulland, C., Laurent, R., Dall'Ara, I., Toupance, B., Chaix, R., 'Genomic Evidence for MHC Disassortative Mating in Humans', *Proceedings of the Royal Society B: Biological Sciences*, 286, 1899 (2019), 20182664.

Daub, J. T., Hofer, T., Cutivet, E., Dupanloup, I., Quintana-Murci, L., Robinson-Rechavi, M., Excoffier, L., 'Evidence for Polygenic Adaptation to Pathogens in the Human Genome', *Molecular Biology and Evolution*, 30, 7 (2013), pp. 1544–1558.

Dausset, J., 'The HLA Adventure', *Transplantation Proceedings*, 31, 1–2 (1999), pp. 22–24.

Degenhardt, J. D., de Candia, P., Chabot, A., Schwartz, S., Henderson, L., Ling, B., Hunter, M., *et al.*, 'Copy Number Variation of *CCL3*-Like Genes Affects Rate of Progression to Simian-AIDS in Rhesus Macaques (*Macaca Mulatta*)', *PLOS Genetics*, 5, 1 (2009), e1000346.

Deschamps, M., Laval, G., Fagny, M., Itan, Y., Abel, L., Casanova, J. L., Patin, E., Quintana-Murci, L., 'Genomic Signatures of Selective Pressures and Introgression from Archaic Hominins at Human Innate Immunity Genes', *American Journal of Human Genetics*, 98, 1 (2016), pp. 5–21.

Elvin, S. J., Williamson, E. D., Scott, J. C., Smith, J. N., Pérez de Lema, G., Chilla, S., Clapham, P., *et al.*, 'Ambiguous Role of CCR5 in *Y. Pestis* Infection', *Nature*, 430, 6998 (2004), p. 418.

Enard, D., Depaulis, F., Roest Crollius, H., 'Human and Non-Human Primate Genomes Share Hotspots of Positive Selection', *PLOS Genetics*, 6, 2 (2010), e1000840.

Fagny, M., Patin, E., Enard, D., Barreiro, L. B., Quintana-Murci, L., Laval, G., 'Exploring the Occurrence of Classic Selective Sweeps in Humans Using Whole-Genome Sequencing Data Sets', *Molecular Biology and Evolution*, 31, 7 (2014), pp. 1850–1868.

Ferrer-Admetlla, A., Sikora, M., Laayouni, H., Esteve, A., Roubinet, F., Blancher, A., Calafell, F., Bertranpetit, J., Casals, F., 'A Natural History of *FUT2* Polymorphism in Humans', *Molecular Biology and Evolution*, 26, 9 (2009), pp. 1993–2003.

Fry, A. E., Ghansa, A., Small, K. S., Palma, A., Auburn, S., Diakite, M., Green, A., *et al.*, 'Positive Selection of a *CD36* Nonsense Variant in Sub-Saharan Africa, but No Association with Severe Malaria Phenotypes', *Human Molecular Genetics*, 18, 14 (2009), pp. 2683–2692.

Fumagalli, M., Cagliani, R., Pozzoli, U., Riva, S., Comi, G. P., Menozzi, G., Bresolin, N., Sironi, M., 'Widespread Balancing Selection and Pathogen-Driven Selection at Blood Group Antigen Genes', *Genome Research*, 19, 2 (2009), pp. 199–212.

Fumagalli, M., Pozzoli, U., Cagliani, R., Comi, G. P., Bresolin, N., Clerici, M., Sironi, M., 'The Landscape of Human Genes Involved in the Immune Response to Parasitic Worms', *BMC Evolutionary Biology*, 10 (2010), p. 264.

Fumagalli, M., Pozzoli, U., Cagliani, R., Comi, G. P., Riva, S., Clerici, M., Bresolin, N., Sironi, M., 'Parasites Represent a Major Selective Force for Interleukin Genes and Shape the Genetic Predisposition to Autoimmune Conditions', *Journal of Experimental Medicine*, 206, 6 (2009), pp. 1395–1408.

Fumagalli, M., Sironi, M., 'Human Genome Variability, Natural Selection and Infectious Diseases', *Current Opinion in Immunology*, 30 (2014), pp. 9–16.

Fumagalli, M., Sironi, M., Pozzoli, U., Ferrer-Admetlla, A., Pattini, L., Nielsen, R., 'Signatures of Environmental Genetic Adaptation Pinpoint Pathogens as the Main Selective Pressure through Human Evolution', *PLOS Genetics*, 7, 11 (2011), e1002355.

Galvani, A. P., Novembre, J., 'The Evolutionary History of the *CCR5-Δ32* HIV-Resistance Mutation', *Microbes and Infection*, 7, 2 (2005), pp. 302–309.

Galvani, A. P., Slatkin, M., 'Evaluating Plague and Smallpox as Historical Selective Pressures for the *CCR5-Δ32* HIV-Resistance Allele', *Proceedings of the National Academy of Sciences USA*, 100, 25 (2003), pp. 15276–15279.

Gomez, F., Hirbo, J., Tishkoff, S. A., 'Genetic Variation and Adaptation in Africa: Implications for Human Evolution and Disease', *Cold Spring Harbor Perspectives in Biology*, 6, 7 (2014), a008524.

Gouy, A., Daub, J. T., Excoffier, L., 'Detecting Gene Subnetworks under Selection in Biological Pathways', *Nucleic Acids Research*, 45, 16 (2017), e149.

Gouy, A., Excoffier, L., 'Polygenic Patterns of Adaptive Introgression in Modern Humans Are Mainly Shaped by Response to Pathogens', *Molecular Biology and Evolution*, 37, 5 (2020), pp. 1420–1433.

Gunalan, K., Lo, E., Hostetler, J. B., Yewhalaw, D., Mu, J., Neafsey, D. E., Yan, G., Miller, L. H., 'Role of *Plasmodium Vivax* Duffy-Binding Protein 1 in Invasion of Duffy-Null Africans', *Proceedings of the National Academy of Sciences USA*, 113, 22 (2016), pp. 6271–6276.

Hahn, M. W., Demuth, J. P., Han, S. G., 'Accelerated Rate of Gene Gain and Loss in Primates', *Genetics*, 177, 3 (2007), pp. 1941–1949.

Haldane, J. B. S., 'Disease and Evolution', *Rice Science*, 19 (Suppl. A) (1949), pp. 68–76.

Hamblin, M. T., Thompson, E. E., Di Rienzo, A., 'Complex Signatures of Natural Selection at the Duffy Blood Group Locus', *American Journal of Human Genetics*, 70, 2 (2002), pp. 369–383.

Hill, A. V., 'Aspects of Genetic Susceptibility to Human Infectious Diseases', *Annual Review of Genetics*, 40 (2006), pp. 469–486.

Hodgson, J. A., Pickrell, J. K., Pearson, L. N., Quillen, E. E., Prista, A., Rocha, J., Soodyall, H., Shriver, M. D., Perry, G. H., 'Natural Selection for the Duffy-Null Allele in the Recently Admixed People of Madagascar', *Proceedings of the Royal Society B: Biological Sciences*, 281, 1789 (2014), 20140930.

Hutagalung, R., Wilairatana, P., Looareesuwan, S., Brittenham, G. M., Aikawa, M., Gordeuk, V. R., 'Influence of Hemoglobin E Trait on the Severity of Falciparum Malaria', *Journal of Infectious Diseases*, 179, 1 (1999), pp. 283–286.

Idaghdour, Y., Quinlan, J., Goulet, J. P., Berghout, J., Gbeha, E., Bruat, V., de Malliard, T., *et al.*, 'Evidence for Additive and Interaction Effects of Host Genotype and Infection in Malaria', *Proceedings of the National Academy of Sciences USA*, 109, 42 (2012), pp. 16786–16793.

Jallow, M., Teo, Y. Y., Small, K. S., Rockett, K. A., Deloukas, P., Clark, T. G., Kivinen, K., *et al.*, 'Genome-Wide and Fine-Resolution Association Analysis of Malaria in West Africa', *Nature Genetics*, 41, 6 (2009), pp. 657–665.

Jin, W., Xu, S., Wang, H., Yu, Y., Shen, Y., Wu, B., Jin, L., 'Genome-Wide Detection of Natural Selection in African Americans Pre- and Post-Admixture', *Genome Research*, 22, 3 (2012), pp. 519–527.

Joy, D. A., Feng, X., Mu, J., Furuya, T., Chotivanich, K., Krettli, A. U., Ho, M., *et al.*, 'Early Origin and Recent Expansion of *Plasmodium Falciparum*', *Science*, 300, 5617 (2003), pp. 318–321.

Karlsson, E. K., Harris, J. B., Tabrizi, S., Rahman, A., Shlyakhter, I., Patterson, N., O'Dushlaine, C., *et al.*, 'Natural Selection in a Bangladeshi Population from the Cholera-Endemic Ganges River Delta', *Science Translational Medicine*, 5, 192 (2013), 192ra86.

Karlsson, E. K., Kwiatkowski, D. P., Sabeti, P. C., 'Natural Selection and Infectious Disease in Human Populations', *Nature Review Genetics*, 15, 6 (2014), pp. 379–393.

Keele, B. F., Jones, J. H., Terio, K. A., Estes, J. D., Rudicell, R. S., Wilson, M. L., Li, Y., *et al.*, 'Increased Mortality and AIDS-Like Immunopathology in Wild Chimpanzees Infected with SIVcpz', *Nature*, 460, 7254 (2009), pp. 515–519.

Kerner, G., Laval, G., Patin, E., Boisson-Dupuis, S., Abel, L., Casanova, J. L., Quintana-Murci, L., 'Human Ancient DNA Analyses Reveal the High Burden of Tuberculosis in Europeans over the Last 2,000 Years', *American Journal of Human Genetics*, 108, 3 (2021), pp. 517–524.

Kerner, G., Patin, E., Quintana-Murci, L., 'New Insights into Human Immunity from Ancient Genomics', *Current Opinion in Immunology*, 72 (2021), pp. 116–125.

Kerner, G., Ramirez-Alejo, N., Seeleuthner, Y., Yang, R., Ogishi, M., Cobat, A., Patin, E., *et al.*, 'Homozygosity for *TYK2* P1104A Underlies Tuberculosis in about 1% of Patients in a Cohort of European Ancestry', *Proceedings of the National Academy of Sciences USA*, 116, 21 (2019), pp. 10430–10434.

Klein, J., Satta, Y., O'Huigin, C., Takahata, N., 'The Molecular Descent of the Major Histocompatibility Complex', *Annual Review of Immunology*, 11 (1993), pp. 269–295.

Koda, Y., Tachida, H., Pang, H., Liu, Y., Soejima, M., Ghaderi, A. A., Takenaka, O., Kimura, H., 'Contrasting Patterns of Polymorphisms at the ABO-Secretor Gene (*FUT2*) and Plasma α(1,3)Fucosyltransferase Gene (*FUT6*) in Human Populations', *Genetics*, 158, 2 (2001), pp. 747–756.

Kulkarni, S., Martin, M. P., Carrington, M., 'The Yin and Yang of HLA and KIR in Human Disease', *Seminars in Immunology*, 20, 6 (2008), pp. 343–352.

Kwiatkowski, D. P., 'How Malaria Has Affected the Human Genome and What Human Genetics Can Teach Us about Malaria', *American Journal of Human Genetics*, 77, 2 (2005), pp. 171–192.

Laayouni, H., Oosting, M., Luisi, P., Ioana, M., Alonso, S., Ricano-Ponce, I., Trynka, G., *et al.*, 'Convergent Evolution in European and Roma Populations Reveals Pressure Exerted by Plague on Toll-Like

Receptors', *Proceedings of the National Academy of Sciences USA*, 111, 7 (2014), pp. 2668–2673.

Lalani, A. S., Masters, J., Zeng, W., Barrett, J., Pannu, R., Everett, H., Arendt, C. W., McFadden, G., 'Use of Chemokine Receptors by Poxviruses', *Science*, 286, 5446 (1999), pp. 1968–1971.

Laso-Jadart, R., Harmant, C., Quach, H., Zidane, N., Tyler-Smith, C., Mehdi, Q., Ayub, Q., Quintana-Murci, L., Patin, E., 'The Genetic Legacy of the Indian Ocean Slave Trade: Recent Admixture and Post-Admixture Selection in the Makranis of Pakistan', *American Journal of Human Genetics*, 101, 6 (2017), pp. 977–984.

Laval, G., Peyregne, S., Zidane, N., Harmant, C., Renaud, F., Patin, E., Prugnolle, F., Quintana-Murci, L., 'Recent Adaptive Acquisition by African Rainforest Hunter-Gatherers of the Late Pleistocene Sickle-Cell Mutation Suggests Past Differences in Malaria Exposure', *American Journal of Human Genetics*, 104, 3 (2019), pp. 553–561.

Leffler, E. M., Gao, Z., Pfeifer, S., Ségurel, L., Auton, A., Venn, O., Bowden, R., *et al.*, 'Multiple Instances of Ancient Balancing Selection Shared between Humans and Chimpanzees', *Science*, 339, 6127 (2013), pp. 1578–1582.

Libert, F., Cochaux, P., Beckman, G., Samson, M., Aksenova, M., Cao, A., Czeizel, A., *et al.*, 'The $\Delta CCR5$ Mutation Conferring Protection against HIV-1 in Caucasian Populations Has a Single and Recent Origin in Northeastern Europe', *Human Molecular Genetics*, 7, 3 (1998), pp. 399–406.

Lindesmith, L., Moe, C., Marionneau, S., Ruvoen, N., Jiang, X., Lindblad, L., Stewart, P., LePendu, J., Baric, R., 'Human Susceptibility and Resistance to Norwalk Virus Infection', *Nature Medicine*, 9, 5 (2003), pp. 548–553.

Lindo, J., Huerta-Sánchez, E., Nakagome, S., Rasmussen, M., Petzelt, B., Mitchell, J., Cybulski, J. S., *et al.*, 'A Time Transect of Exomes from a Native American Population before and after European Contact', *Nature Communications*, 7 (2016), pp. 13175.

Livingstone, F. B., 'The Duffy Blood Groups, Vivax Malaria, and Malaria Selection in Human Populations: A Review', *Human Biology*, 56, 3 (1984), pp. 413–425.

Lopez, M., Choin, J., Sikora, M., Siddle, K., Harmant, C., Costa, H. A., Silvert, M., *et al.*, 'Genomic Evidence for Local Adaptation of Hunter-Gatherers to the African Rainforest', *Current Biology*, 29, 17 (2019), pp. 2926–2935, e4.

Louicharoen, C., Patin, E., Paul, R., Nuchprayoon, I., Witoonpanich, B., Peerapittayamongkol, C., Casademont, I., *et al.*, 'Positively Selected *G6PD*-Mahidol Mutation Reduces *Plasmodium Vivax* Density in Southeast Asians', *Science*, 326, 5959 (2009), pp. 1546–1549.

MacArthur, D. G., Balasubramanian, S., Frankish, A., Huang, N., Morris, J., Walter, K., Jostins, L., *et al.*, 'A Systematic Survey of Loss-of-Function Variants in Human Protein-Coding Genes', *Science*, 335, 6070 (2012), pp. 823–828.

Malaria Genomic Epidemiology Network, 'Reappraisal of Known Malaria Resistance Loci in a Large Multicenter Study', *Nature Genetics*, 46, 11 (2014), pp. 1197–1204.

Manry, J., Laval, G., Patin, E., Fornarino, S., Itan, Y., Fumagalli, M., Sironi, M., *et al.*, 'Evolutionary Genetic Dissection of Human Interferons', *Journal of Experimental Medicine*, 208, 13 (2011), pp. 2747–2759.

Martin, M. P., Dean, M., Smith, M. W., Winkler, C., Gerrard, B., Michael, N. L., Lee, B., *et al.*, 'Genetic Acceleration of AIDS Progression by a Promoter Variant of *CCR5*', *Science*, 282, 5395 (1998), pp. 1907–1911.

Mathieson, I., Lazaridis, I., Rohland, N., Mallick, S., Patterson, N., Roodenberg, S. A., Harney, E., *et al.*, 'Genome-Wide Patterns of Selection in 230 Ancient Eurasians', *Nature*, 528, 7583 (2015), pp. 499–503.

McManus, K. F., Taravella, A. M., Henn, B. M., Bustamante, C. D., Sikora, M., Cornejo, O. E., 'Population Genetic Analysis of the DARC Locus (Duffy) Reveals Adaptation from Standing Variation Associated with Malaria Resistance in Humans', *PLOS Genetics*, 13, 3 (2017), e1006560.

Mecsas, J., Franklin, G., Kuziel, W. A., Brubaker, R. R., Falkow, S., Mosier, D. E., 'CCR5 Mutation and Plague Protection', *Nature*, 427, 6975 (2004), p. 606.

Menard, D., Barnadas, C., Bouchier, C., Henry-Halldin, C., Gray, L. R., Ratsimbasoa, A., Thonier, V., *et al.*, '*Plasmodium Vivax* Clinical Malaria Is Commonly Observed in Duffy-Negative Malagasy People',

Proceedings of the National Academy of Sciences USA, 107, 13 (2010), pp. 5967–5971.

Michon, P., Woolley, I., Wood, E. M., Kastens, W., Zimmerman, P. A., Adams, J. H., 'Duffy-Null Promoter Heterozygosity Reduces DARC Expression and Abrogates Adhesion of the *P. Vivax* Ligand Required for Blood-Stage Infection', *FEBS Letters*, 495, 1–2 (2001), pp. 111–114.

Modiano, D., Luoni, G., Sirima, B. S., Simpore, J., Verra, F., Konate, A., Rastrelli, E., *et al.*, 'Haemoglobin C Protects against Clinical *Plasmodium Falciparum* Malaria', *Nature*, 414, 6861 (2001), pp. 305–308.

Morens, D. M., Folkers, G. K., Fauci, A. S., 'Emerging Infections: A Perpetual Challenge', *The Lancet Infectious Diseases*, 8, 11 (2008), pp. 710–719.

Novembre, J., Galvani, A. P., Slatkin, M., 'The Geographic Spread of the CCR5 Δ32 HIV-Resistance Allele', *PLOS Biology*, 3, 11 (2005), e339.

Ober, C., Weitkamp, L. R., Cox, N., Dytch, H., Kostyu, D., Elias, S., 'HLA and Mate Choice in Humans', *American Journal of Human Genetics*, 61, 3 (1997), pp. 497–504.

O'Brien, S. J., Moore, J. P., 'The Effect of Genetic Variation in Chemokines and Their Receptors on HIV Transmission and Progression to AIDS', *Immunology Review*, 177 (2000), pp. 99–111.

Ohashi, J., Naka, I., Patarapotikul, J., Hananantachai, H., Brittenham, G., Looareesuwan, S., Clark, A. G., Tokunaga, K., 'Extended Linkage Disequilibrium Surrounding the Hemoglobin E Variant Due to Malarial Selection', *American Journal of Human Genetics*, 74, 6 (2004), pp. 1198–1208.

Olalde, I., Allentoft, M. E., Sanchez-Quinto, F., Santpere, G., Chiang, C. W., DeGiorgio, M., Prado-Martinez, J., *et al.*, 'Derived Immune and Ancestral Pigmentation Alleles in a 7,000-Year-Old Mesolithic European', *Nature*, 507, 7491 (2014), pp. 225–228.

Olson, M. V., 'When Less Is More: Gene Loss as an Engine of Evolutionary Change', *American Journal of Human Genetics*, 64, 1 (1999), pp. 18–23.

Parkes, M., Cortes, A., Van Heel, D. A., Brown, M. A., 'Genetic Insights into Common Pathways and Complex Relationships among Immune-Mediated Diseases', *Nature Review Genetics*, 14, 9 (2013), pp. 661–673.

Perry, G. H., Yang, F., Marques-Bonet, T., Murphy, C., Fitzgerald, T., Lee, A. S., Hyland, C., *et al.*, 'Copy Number Variation and Evolution in Humans and Chimpanzees', *Genome Research*, 18, 11 (2008), pp. 1698–1710.

Pierron, D., Heiske, M., Razafindrazaka, H., Pereda-Loth, V., Sanchez, J., Alva, O., Arachiche, A., *et al.*, 'Strong Selection during the Last Millennium for African Ancestry in the Admixed Population of Madagascar', *Nature Communications*, 9, 1 (2018), p. 932.

Prugnolle, F., Manica, A., Charpentier, M., Guegan, J. F., Guernier, V., Balloux, F., 'Pathogen-Driven Selection and Worldwide HLA Class I Diversity', *Current Biology*, 15, 11 (2005), pp. 1022–1027.

Quach, H., Quintana-Murci, L., 'Living in an Adaptive World: Genomic Dissection of the Genus *Homo* and Its Immune Response', *Journal of Experimental Medicine*, 214, 4 (2017), pp. 877–894.

Quintana-Murci, L., 'Human Immunology through the Lens of Evolutionary Genetics', *Cell*, 177, 1 (2019), pp. 184–199.

Quintana-Murci, L., Alcaïs, A., Abel, L., Casanova, J. L., 'Immunology *in Natura*: Clinical, Epidemiological and Evolutionary Genetics of Infectious Diseases', *Nature Immunology*, 8, 11 (2007), pp. 1165–1171.

Quintana-Murci, L., Barreiro, L. B., 'The Role Played by Natural Selection on Mendelian Traits in Humans', *Annals of the New York Academy of Sciences*, 1214 (2010), pp. 1–17.

Quintana-Murci, L., Clark, A. G., 'Population Genetic Tools for Dissecting Innate Immunity in Humans', *Nature Reviews Immunology*, 13, 4 (2013), pp. 280–293.

Rajagopalan, S., Long, E. O., 'Understanding How Combinations of HLA and KIR Genes Influence Disease', *Journal of Experimental Medicine*, 201, 7 (2005), pp. 1025–1029.

Ruwende, C., Hill, A., 'Glucose-6-Phosphate Dehydrogenase Deficiency and Malaria', *Journal of Molecular Medicine*, 76, 8 (1998), pp. 581–588.

Ruwende, C., Khoo, S. C., Snow, R. W., Yates, S. N., Kwiatkowski, D., Gupta, S., Warn, P., *et al.*, 'Natural Selection of Hemi- and Heterozygotes for G6PD Deficiency in Africa by Resistance to Severe Malaria', *Nature*, 376, 6537 (1995), pp. 246–249.

Ryan, J. R., Stoute, J. A., Amon, J., Dunton, R. F., Mtalib, R., Koros, J., Owour, B., *et al.*, 'Evidence for Transmission of *Plasmodium Vivax* among a Duffy Antigen Negative Population in Western Kenya', *American Journal of Tropical Medicine and Hygiene*, 75, 4 (2006), pp. 575–581.

Sabeti, P., Usen, S., Farhadian, S., Jallow, M., Doherty, T., Newport, M., Pinder, M., Ward, R., Kwiatkowski, D., 'CD40l Association with Protection from Severe Malaria', *Genes & Immunity*, 3, 5 (2002), pp. 286–291.

Sabeti, P. C., Walsh, E., Schaffner, S. F., Varilly, P., Fry, B., Hutcheson, H. B., Cullen, M., *et al.*, 'The Case for Selection at *CCR5-Δ32*', *PLOS Biology*, 3, 11 (2005), e378.

Sanz, J., Randolph, H. E., Barreiro, L. B., 'Genetic and Evolutionary Determinants of Human Population Variation in Immune Responses', *Current Opinion in Genetics & Development*, 53 (2018), pp. 28–35.

Ségurel, L., Gao, Z., Przeworski, M., 'Ancestry Runs Deeper than Blood: The Evolutionary History of ABO Points to Cryptic Variation of Functional Importance', *Bioessays*, 35, 10 (2013), pp. 862–867.

Ségurel, L., Quintana-Murci, L., 'Preserving Immune Diversity through Ancient Inheritance and Admixture', *Current Opinion in Immunology*, 30 (2014), pp. 79–84.

Ségurel, L., Thompson, E. E., Flutre, T., Lovstad, J., Venkat, A., Margulis, S. W., Moyse, J., *et al.*, 'The ABO Blood Group Is a Trans-Species Polymorphism in Primates', *Proceedings of the National Academy of Sciences USA*, 109, 45 (2012), pp. 18493–18498.

Sharp, P. M., Shaw, G. M., Hahn, B. H., 'Simian Immunodeficiency Virus Infection of Chimpanzees', *Journal of Virology*, 79, 7 (2005), pp. 3891–3902.

Siddle, K. J., Quintana-Murci, L., 'The Red Queen's Long Race: Human Adaptation to Pathogen Pressure', *Current Opinion in Genetics & Development*, 29 (2014), pp. 31–38.

Single, R. M., Martin, M. P., Gao, X., Meyer, D., Yeager, M., Kidd, J. R., Kidd, K. K., Carrington, M., 'Global Diversity and Evidence for Coevolution of KIR and HLA', *Nature Genetics*, 39, 9 (2007), pp. 1114–1119.

Stephens, J. C., Reich, D. E., Goldstein, D. B., Shin, H. D., Smith, M. W., Carrington, M., Winkler, C., *et al.*, 'Dating the Origin of the *CCR5-Δ32*

AIDS-Resistance Allele by the Coalescence of Haplotypes', *American Journal of Human Genetics*, 62, 6 (1998), pp. 1507–1515.

Thorven, M., Grahn, A., Hedlund, K.-O., Johansson, H., Wahlfrid, C., Larson, G., Svensson, L., 'A Homozygous Nonsense Mutation (428G→A) in the Human Secretor (*FUT2*) Gene Provides Resistance to Symptomatic Norovirus (*GGII*) Infections', *Journal of Virology*, 79, 24 (2005), pp. 15351–15355.

Tishkoff, S. A., Varkonyi, R., Cahinhinan, N., Abbes, S., Argyropoulos, G., Destro-Bisol, G., Drousiotou, A., *et al.*, 'Haplotype Diversity and Linkage Disequilibrium at Human *G6PD*: Recent Origin of Alleles that Confer Malarial Resistance', *Science*, 293, 5529 (2001), pp. 455–462.

Tournamille, C., Colin, Y., Cartron, J. P., Le Van Kim, C., 'Disruption of a GATA Motif in the Duffy Gene Promoter Abolishes Erythroid Gene Expression in Duffy-Negative Individuals', *Nature Genetics*, 10, 2 (1995), pp. 224–228.

Van Valen, L., 'A New Evolutionary Law', *Evolutionary Theory*, 1 (1973), pp. 1–30.

Varki, A., 'A Chimpanzee Genome Project Is a Biomedical Imperative', *Genome Research*, 10, 8 (2000), pp. 1065–1070.

Varki, A., Altheide, T. K., 'Comparing the Human and Chimpanzee Genomes: Searching for Needles in a Haystack', *Genome Research*, 15, 12 (2005), pp. 1746–1758.

Vasseur, E., Boniotto, M., Patin, E., Laval, G., Quach, H., Manry, J., Crouau-Roy, B., Quintana-Murci, L., 'The Evolutionary Landscape of Cytosolic Microbial Sensors in Humans', *American Journal of Human Genetics*, 91, 1 (2012), pp. 27–37.

Vasseur, E., Patin, E., Laval, G., Pajon, S., Fornarino, S., Crouau-Roy, B., Quintana-Murci, L., 'The Selective Footprints of Viral Pressures at the Human RIG-I-Like Receptor Family', *Human Molecular Genetics*, 20, 22 (2011), pp. 4462–4474.

Verdu, P., Barreiro, L. B., Patin, E., Gessain, A., Cassar, O., Kidd, J. R., Kidd, K. K., *et al.*, 'Evolutionary Insights into the High Worldwide Prevalence of *MBL2* Deficiency Alleles', *Human Molecular Genetics*, 15, 17 (2006), pp. 2650–2658.

Walsh, E. C., Sabeti, P., Hutcheson, H. B., Fry, B., Schaffner, S. F., de Bakker, P. I., Varilly, P., *et al.*, 'Searching for Signals of Evolutionary Selection in 168 Genes Related to Immune Function', *Human Genetics*, 119, 1–2 (2006), pp. 92–102.

Wang, X., Grus, W. E., Zhang, J., 'Gene Losses during Human Origins', *PLOS Biology*, 4, 3 (2006), e52.

Wedekind, C., Seebeck, T., Bettens, F., Paepke, A. J., 'MHC-Dependent Mate Preferences in Humans', *Proceedings of the Royal Society B: Biological Sciences*, 260, 1359 (1995), pp. 245–249.

Wooding, S., Stone, A. C., Dunn, D. M., Mummidi, S., Jorde, L. B., Weiss, R. K., Ahuja, S., Bamshad, M. J., 'Contrasting Effects of Natural Selection on Human and Chimpanzee CC Chemokine Receptor 5', *American Journal of Human Genetics*, 76, 2 (2005), pp. 291–301.

Xue, Y., Daly, A., Yngvadottir, B., Liu, M., Coop, G., Kim, Y., Sabeti, P., *et al.*, 'Spread of an Inactive Form of Caspase-12 in Humans Is Due to Recent Positive Selection', *American Journal of Human Genetics*, 78, 4 (2006), pp. 659–670.

Zhernakova, A., Elbers, C. C., Ferwerda, B., Romanos, J., Trynka, G., Dubois, P. C., de Kovel, C. G., *et al.*, 'Evolutionary and Functional Analysis of Celiac Risk Loci Reveals SH2B3 as a Protective Factor against Bacterial Infection', *American Journal of Human Genetics*, 86, 6 (2010), pp. 970–977.

5. Admixture, Culture and Medicine

Abi-Rached, L., Jobin, M. J., Kulkarni, S., McWhinnie, A., Dalva, K., Gragert, L., Babrzadeh, F., *et al.*, 'The Shaping of Modern Human Immune Systems by Multiregional Admixture with Archaic Humans', *Science*, 334, 6052 (2011), pp. 89–94.

Adhikari, K., Chacón-Duque, J. C., Mendoza-Revilla, J., Fuentes-Guajardo, M., Ruiz-Linares, A., 'The Genetic Diversity of the Americas', *Annual Review of Genomics and Human Genetics*, 18 (2017), pp. 277–296.

Aime, C., Laval, G., Patin, E., Verdu, P., Ségurel, L., Chaix, R., Hegay, T., *et al.*, 'Human Genetic Data Reveal Contrasting Demographic Patterns between Sedentary and Nomadic Populations that Predate the Emergence of Farming', *Molecular Biology and Evolution*, 30, 12 (2013), pp. 2629–2644.

Albert, F. W., Kruglyak, L., 'The Role of Regulatory Variation in Complex Traits and Disease', *Nature Review Genetics*, 16, 4 (2015), pp. 197–212.

Ammerman, A. J., Cavalli-Sforza, L. L., *The Neolithic Transition and the Genetics of Populations in Europe*, Princeton, Princeton University Press, 1984.

Bach, J. F., 'The Effect of Infections on Susceptibility to Autoimmune and Allergic Diseases', *New England Journal of Medicine*, 347, 12 (2002), pp. 911–920.

Bamshad, M. J., Watkins, W. S., Dixon, M. E., Jorde, L. B., Rao, B. B., Naidu, J. M., Prasad, B. V., Rasanayagam, A., Hammer, M. F., 'Female Gene Flow Stratifies Hindu Castes', *Nature*, 395, 6703 (1998), pp. 651–652.

Bastard, P., Rosen, L. B., Zhang, Q., Michailidis, E., Hoffmann, H. H., Zhang, Y., Dorgham, K., *et al.*, 'Autoantibodies against Type I IFNs in Patients with Life-Threatening COVID-19', *Science*, 370, eabd4585 (2020).

Bell, J. T., Pai, A. A., Pickrell, J. K., Gaffney, D. J., Pique-Regi, R., Degner, J. F., Gilad, Y., Pritchard, J. K., 'DNA Methylation Patterns Associate with Genetic and Gene Expression Variation in Hapmap Cell Lines', *Genome Biology*, 12, 1 (2011), R10.

Bergman, Y., Cedar, H., 'DNA Methylation Dynamics in Health and Disease', *Nature Structural and Molecular Biology*, 20, 3 (2013), pp. 274–281.

Bonder, M. J., Luijk, R., Zhernakova, D. V., Moed, M., Deelen, P., Vermaat, M., Van Iterson, M., *et al.*, 'Disease Variants Alter Transcription Factor Levels and Methylation of Their Binding Sites', *Nature Genetics*, 49, 1 (2017), pp. 131–138.

Browning, S. R., Browning, B. L., Zhou, Y., Tucci, S., Akey, J. M., 'Analysis of Human Sequence Data Reveals Two Pulses of Archaic Denisovan Admixture', *Cell*, 173, 1 (2018), pp. 53–61, e9.

Bryc, K., Auton, A., Nelson, M. R., Oksenberg, J. R., Hauser, S. L., Williams, S., Froment, A., *et al.*, 'Genome-Wide Patterns of Population Structure and Admixture in West Africans and African Americans', *Proceedings of the National Academy of Sciences USA*, 107, 2 (2010), pp. 786–791.

Busby, G. B., Band, G., Si Le, Q., Jallow, M., Bougama, E., Mangano, V. D., Amenga-Etego, L. N., *et al.*, 'Admixture into and within Sub-Saharan Africa', *eLife*, 5 (2016), e15266.

Buzbas, E. O., Verdu, P., 'Inference on Admixture Fractions in a Mechanistic Model of Recurrent Admixture', *Theoretical Population Biology*, 122 (2018), pp. 149–157.

Carja, O., MacIsaac, J. L., Mah, S. M., Henn, B. M., Kobor, M. S., Feldman, M. W., Fraser, H. B., 'Worldwide Patterns of Human Epigenetic Variation', *Nature Ecology & Evolution*, 1, 10 (2017), pp. 1577–1583.

Cavalli, G., Heard, E., 'Advances in Epigenetics Link Genetics to the Environment and Disease', *Nature*, 571, 7766 (2019), pp. 489–499.

Cavalli-Sforza, L. L., Menozzi, P., Piazza, A., *The History and Geography of Human Genes*, Princeton, Princeton University Press, 1994.

Chacón-Duque, J. C., Adhikari, K., Avendano, E., Campo, O., Ramirez, R., Rojas, W., Ruiz-Linares, A., Restrepo, B. N., Bedoya, G., 'African Genetic Ancestry Is Associated with a Protective Effect on Dengue Severity in Colombian Populations', *Infection, Genetics and Evolution*, 27 (2014), pp. 89–95.

Chaix, R., Austerlitz, F., Khegay, T., Jacquesson, S., Hammer, M. F., Heyer, E., Quintana-Murci, L., 'The Genetic or Mythical Ancestry of Descent Groups: Lessons from the Y Chromosome', *American Journal of Human Genetics*, 75, 6 (2004), pp. 1113–1116.

Chen, L., Ge, B., Casale, F. P., Vasquez, L., Kwan, T., Garrido-Martin, D., Watt, S., *et al.*, 'Genetic Drivers of Epigenetic and Transcriptional Variation in Human Immune Cells', *Cell*, 167, 5 (2016), pp. 1398–1414, e24.

Choin, J., Mendoza-Revilla, J., Arauna, L. R., Cuadros-Espinoza, S., Cassar, O., Larena, M., Ko, A. M., *et al.*, 'Genomic Insights into Population History and Biological Adaptation in Oceania', *Nature*, 592, 7855 (2021), pp. 583–589.

Dannemann, M., Andrés, A. M., Kelso, J., 'Introgression of Neandertal-and Denisovan-Like Haplotypes Contributes to Adaptive Variation in Human Toll-Like Receptors', *American Journal of Human Genetics*, 98, 1 (2016), pp. 22–33.

Dannemann, M., Kelso, J., 'The Contribution of Neanderthals to Phenotypic Variation in Modern Humans', *American Journal of Human Genetics*, 101, 4 (2017), pp. 578–589.

Dannemann, M., Prüfer, K., Kelso, J., 'Functional Implications of Neandertal Introgression in Modern Humans', *Genome Biology*, 18, 1 (2017), p. 61.

Dannemann, M., Racimo, F., 'Something Old, Something Borrowed: Admixture and Adaptation in Human Evolution', *Current Opinion in Genetics & Development*, 53 (2018), pp. 1–8.

Dermitzakis, E. T., 'Cellular Genomics for Complex Traits', *Nature Review Genetics*, 13, 3 (2012), pp. 215–220.

Deschamps, M., Laval, G., Fagny, M., Itan, Y., Abel, L., Casanova, J. L., Patin, E., Quintana-Murci, L., 'Genomic Signatures of Selective Pressures and Introgression from Archaic Hominins at Human Innate Immunity Genes', *American Journal of Human Genetics*, 98, 1 (2016), pp. 5–21.

Dobzhansky, T. H., 'Nothing in Biology Makes Sense Except in the Light of Evolution', *American Biology Teacher*, 35 (1973), pp. 125–129.

Duffy, D., Rouilly, V., Libri, V., Hasan, M., Beitz, B., David, M., Urrutia, A., *et al.*, 'Functional Analysis via Standardized Whole-Blood Stimulation Systems Defines the Boundaries of a Healthy Immune Response to Complex Stimuli', *Immunity*, 40, 3 (2014), pp. 436–450.

Enard, D., Cai, L., Gwennap, C., Petrov, D. A., 'Viruses Are a Dominant Driver of Protein Adaptation in Mammals', *eLife*, 5 (2016), e12469.

Enard, D., Petrov, D. A., 'Evidence that RNA Viruses Drove Adaptive Introgression between Neanderthals and Modern Humans', *Cell*, 175, 2 (2018), pp. 360–371, e13.

Fagny, M., Patin, E., MacIsaac, J. L., Rotival, M., Flutre, T., Jones, M. J., Siddle, K. J., *et al.*, 'The Epigenomic Landscape of African Rainforest Hunter-Gatherers and Farmers', *Nature Communications*, 6 (2015), p. 10047.

Fairfax, B. P., Humburg, P., Makino, S., Naranbhai, V., Wong, D., Lau, E., Jostins, L., *et al.*, 'Innate Immune Activity Conditions the Effect of Regulatory Variants upon Monocyte Gene Expression', *Science*, 343, 6175 (2014), 1246949.

Fairfax, B. P., Knight, J. C., 'Genetics of Gene Expression in Immunity to Infection', *Current Opinion in Immunology*, 30 (2014), pp. 63–71.

Feil, R., Fraga, M. F., 'Epigenetics and the Environment: Emerging Patterns and Implications', *Nature Review Genetics*, 13, 2 (2011), pp. 97–109.

Fortes-Lima, C., Laurent, R., Thouzeau, V., Toupance, B., Verdu, P., 'Complex Genetic Admixture Histories Reconstructed with Approximate Bayesian Computations', *Molecular Ecology Resources*, 21 (2021), pp. 1098–1117.

Fraser, H. B., Lam, L. L., Neumann, S. M., Kobor, M. S., 'Population-Specificity of Human DNA Methylation', *Genome Biology*, 13, 2 (2012), R8.

Gat-Viks, I., Chevrier, N., Wilentzik, R., Eisenhaure, T., Raychowdhury, R., Steuerman, Y., Shalek, A. K., *et al.*, 'Deciphering Molecular Circuits from Genetic Variation Underlying Transcriptional Responsiveness to Stimuli', *Nature Biotechnology*, 31, 4 (2013), pp. 342–349.

Gittelman, R. M., Schraiber, J. G., Vernot, B., Mikacenic, C., Wurfel, M. M., Akey, J. M., 'Archaic Hominin Admixture Facilitated Adaptation to Out-of-Africa Environments', *Current Biology*, 26, 24 (2016), pp. 3375–3382.

Gopalan, S., Carja, O., Fagny, M., Patin, E., Myrick, J. W., McEwen, L. M., Mah, S. M., *et al.*, 'Trends in DNA Methylation with Age Replicate across Diverse Human Populations', *Genetics*, 206, 3 (2017), pp. 1659–1674.

Gosling, A. L., Buckley, H. R., Matisoo-Smith, E., Merriman, T. R., 'Pacific Populations, Metabolic Disease and "Just-So Stories": A Critique of the "Thrifty Genotype" Hypothesis in Oceania', *Annals of Human Genetics*, 79, 6 (2015), pp. 470–480.

Gosling, A. L., Matisoo-Smith, E., Merriman, T. R., 'Hyperuricaemia in the Pacific: Why the Elevated Serum Urate Levels?', *Rheumatology International*, 34, 6 (2014), pp. 743–757.

Haber, M., Doumet-Serhal, C., Scheib, C. L., Xue, Y., Mikulski, R., Martiniano, R., Fischer-Genz, B., *et al.*, 'A Transient Pulse of Genetic Admixture from the Crusaders in the Near East Identified from Ancient Genome Sequences', *American Journal of Human Genetics*, 104, 5 (2019), pp. 977–984.

Hammer, M. F., Mendez, F. L., Cox, M. P., Woerner, A. E., Wall, J. D., 'Sex-Biased Evolutionary Forces Shape Genomic Patterns of Human Diversity', *PLOS Genetics*, 4, 9 (2008), e1000202.

Harris, K., Nielsen, R., 'The Genetic Cost of Neanderthal Introgression', *Genetics*, 203, 2 (2016), pp. 881–891.

Harrison, G. F., Sanz, J., Boulais, J., Mina, M. J., Grenier, J.-C., Leng, Y., Dumaine, A., *et al.*, 'Natural Selection Contributed to Immunological Differences between Hunter-Gatherers and Agriculturalists', *Nature Ecology & Evolution*, 3, 8 (2019), pp. 1253–1264.

Hedrick, P. W., 'Adaptive Introgression in Animals: Examples and Comparison to New Mutation and Standing Variation as Sources of Adaptive Variation', *Molecular Ecology*, 22, 18 (2013), pp. 4606–4618.

Hellenthal, G., Busby, G. B., Band, G., Wilson, J. F., Capelli, C., Falush, D., Myers, S., 'A Genetic Atlas of Human Admixture History', *Science*, 343, 6172 (2014), pp. 747–751.

Henn, B. M., Cavalli-Sforza, L. L., Feldman, M. W., 'The Great Human Expansion', *Proceedings of the National Academy of Sciences USA*, 109, 44 (2012), pp. 17758–17764.

Heyer, E., Chaix, R., Pavard, S., Austerlitz, F., 'Sex-Specific Demographic Behaviours that Shape Human Genomic Variation', *Molecular Ecology*, 21, 3 (2012), pp. 597–612.

Heyn, H., Esteller, M., 'DNA Methylation Profiling in the Clinic: Applications and Challenges', *Nature Review Genetics*, 13, 10 (2012), pp. 679–692.

Heyn, H., Moran, S., Hernando-Herraez, I., Sayols, S., Gomez, A., Sandoval, J., Monk, D., *et al.*, 'DNA Methylation Contributes to Natural Human Variation', *Genome Research*, 23, 9 (2013), pp. 1363–1372.

Hodgson, J. A., Pickrell, J. K., Pearson, L. N., Quillen, E. E., Prista, A., Rocha, J., Soodyall, H., Shriver, M. D., Perry, G. H., 'Natural Selection for the Duffy-Null Allele in the Recently Admixed People of Madagascar', *Proceedings of the Royal Society B: Biological Sciences*, 281, 1789 (2014), 20140930.

Horvath, S., Gurven, M., Levine, M. E., Trumble, B. C., Kaplan, H., Allayee, H., Ritz, B. R., *et al.*, 'An Epigenetic Clock Analysis of Race/Ethnicity, Sex, and Coronary Heart Disease', *Genome Biology*, 17, 1 (2016), p. 171.

Hudjashov, G., Karafet, T. M., Lawson, D. J., Downey, S., Savina, O., Sudoyo, H., Lansing, J. S., Hammer, M. F., Cox, M. P., 'Complex Patterns of Admixture across the Indonesian Archipelago', *Molecular Biology and Evolution*, 34, 10 (2017), pp. 2439–2452.

Huerta-Sánchez, E., Jin, X., Asan, Bianba, Z., Peter, B. M., Vinckenbosch, N., Liang, Y., *et al.*, 'Altitude Adaptation in Tibetans Caused by Introgression of Denisovan-Like DNA', *Nature*, 512, 7513 (2014), pp. 194–197.

Husquin, L. T., Rotival, M., Fagny, M., Quach, H., Zidane, N., McEwen, L. M., MacIsaac, J. L., *et al.*, 'Exploring the Genetic Basis of Human Population Differences in DNA Methylation and Their Causal Impact on Immune Gene Regulation', *Genome Biology*, 19, 1 (2018), p. 222.

Idaghdour, Y., Czika, W., Shianna, K. V., Lee, S. H., Visscher, P. M., Martin, H. C., Miclaus, K., *et al.*, 'Geographical Genomics of Human Leukocyte Gene Expression Variation in Southern Morocco', *Nature Genetics*, 42, 1 (2010), pp. 62–67.

Idaghdour, Y., Storey, J. D., Jadallah, S. J., Gibson, G., 'A Genome-Wide Gene Expression Signature of Environmental Geography in Leukocytes of Moroccan Amazighs', *PLOS Genetics*, 4, 4 (2008), e1000052.

Jacobs, G. S., Hudjashov, G., Saag, L., Kusuma, P., Darusallam, C. C., Lawson, D. J., Mondal, M., *et al.*, 'Multiple Deeply Divergent Denisovan Ancestries in Papuans', *Cell*, 177, 4 (2019), pp. 1010–1021, e32.

Jeong, C., Balanovsky, O., Lukianova, E., Kahbatkyzy, N., Flegontov, P., Zaporozhchenko, V., Immel, A., *et al.*, 'The Genetic History of Admixture across Inner Eurasia', *Nature Ecology & Evolution*, 3, 6 (2019), pp. 966–976.

Jin, W., Xu, S., Wang, H., Yu, Y., Shen, Y., Wu, B., Jin, L., 'Genome-Wide Detection of Natural Selection in African Americans Pre- and Post-Admixture', *Genome Research*, 22, 3 (2012), pp. 519–527.

Kaminsky, Z. A., Tang, T., Wang, S.-C., Ptak, C., Oh, G. H. T., Wong, A. H. C., Feldcamp, L. A., *et al.*, 'DNA Methylation Profiles in Monozygotic and Dizygotic Twins', *Nature Genetics*, 41, 2 (2009), pp. 240–245.

Kerner, G., Patin, E., Quintana-Murci, L., 'New Insights into Human Immunity from Ancient Genomics', *Current Opinion in Immunology*, 72 (2021), pp. 116–125.

Kim-Hellmuth, S., Bechheim, M., Pütz, B., Mohammadi, P., Nédélec, Y., Giangreco, N., Becker, J., *et al.*, 'Genetic Regulatory Effects Modified by Immune Activation Contribute to Autoimmune Disease Associations', *Nature Communications*, 8, 1 (2017), p. 266.

Kittler, R., Kayser, M., Stoneking, M., 'Molecular Evolution of *Pediculus Humanus* and the Origin of Clothing', *Current Biology*, 13, 16 (2003), pp. 1414–1417.

Kudaravalli, S., Veyrieras, J. B., Stranger, B. E., Dermitzakis, E. T., Pritchard, J. K., 'Gene Expression Levels Are a Target of Recent Natural Selection in the Human Genome', *Molecular Biology and Evolution*, 26, 3 (2009), pp. 649–658.

Kulis, M., Esteller, M., 'DNA Methylation and Cancer', *Advanced Genetics*, 70 (2010), pp. 27–56.

Lam, L. L., Emberly, E., Fraser, H. B., Neumann, S. M., Chen, E., Miller, G. E., Kobor, M. S., 'Factors Underlying Variable DNA Methylation in a Human Community Cohort', *Proceedings of the National Academy of Sciences USA*, 109, Suppl. 2 (2012), pp. 17253–17260.

Laso-Jadart, R., Harmant, C., Quach, H., Zidane, N., Tyler-Smith, C., Mehdi, Q., Ayub, Q., Quintana-Murci, L., Patin, E., 'The Genetic Legacy of the Indian Ocean Slave Trade: Recent Admixture and Post-Admixture Selection in the Makranis of Pakistan', *American Journal of Human Genetics*, 101, 6 (2017), pp. 977–984.

Laval, G., Peyregne, S., Zidane, N., Harmant, C., Renaud, F., Patin, E., Prugnolle, F., Quintana-Murci, L., 'Recent Adaptive Acquisition by African Rainforest Hunter-Gatherers of the Late Pleistocene Sickle-Cell Mutation Suggests Past Differences in Malaria Exposure', *American Journal of Human Genetics*, 104, 3 (2019), pp. 553–561.

Lazaridis, I., Patterson, N., Mittnik, A., Renaud, G., Mallick, S., Kirsanow, K., Sudmant, P. H., *et al.*, 'Ancient Human Genomes Suggest Three Ancestral Populations for Present-Day Europeans', *Nature*, 513, 7518 (2014), pp. 409–413.

Lee, M. N., Ye, C., Villani, A. C., Raj, T., Li, W., Eisenhaure, T. M., Imboywa, S. H., *et al.*, 'Common Genetic Variants Modulate Pathogen-Sensing Responses in Human Dendritic Cells', *Science*, 343, 6175 (2014), 1246980.

Lindo, J., Huerta-Sánchez, E., Nakagome, S., Rasmussen, M., Petzelt, B., Mitchell, J., Cybulski, J. S., *et al.*, 'A Time Transect of Exomes from a Native American Population before and after European Contact', *Nature Communications*, 7 (2016), p. 13175.

Lipson, M., Loh, P.-R., Patterson, N., Moorjani, P., Ko, Y.-C., Stoneking, M., Berger, B., Reich, D., 'Reconstructing Austronesian Population History in Island Southeast Asia', *Nature Communications*, 5 (2014), p. 4689.

Loh, P.-R., Lipson, M., Patterson, N., Moorjani, P., Pickrell, J. K., Reich, D., Berger, B., 'Inferring Admixture Histories of Human Populations Using Linkage Disequilibrium', *Genetics*, 193, 4 (2013), pp. 1233–1254.

Lokki, A. I., Järvelä, I., Israelsson, E., Maiga, B., Troye-Blomberg, M., Dolo, A., Doumbo, O. K., Meri, S., Holmberg, V., 'Lactase Persistence Genotypes and Malaria Susceptibility in Fulani of Mali', *Malaria Journal*, 10 (2011), p. 9.

Marchi, N., Mennecier, P., Georges, M., Lafosse, S., Hegay, T., Dorzhu, C., Chichlo, B., Ségurel, L., Heyer, E., 'Close Inbreeding and Low Genetic Diversity in Inner Asian Human Populations despite Geographical Exogamy', *Scientific Reports*, 8, 1 (2018), p. 9397.

Martin, S. H., Jiggins, C. D., 'Interpreting the Genomic Landscape of Introgression', *Current Opinion in Genetics & Development*, 47 (2017), pp. 69–74.

Matisoo-Smith, E., Gosling, A. L., 'Walking Backwards into the Future: The Need for a Holistic Evolutionary Approach in Pacific Health Research', *Annals of Human Biology*, 45, 3 (2018), pp. 175–187.

McCoy, R. C., Wakefield, J., Akey, J. M., 'Impacts of Neanderthal-Introgressed Sequences on the Landscape of Human Gene Expression', *Cell*, 168, 5 (2017), pp. 916–927, e12.

Mendez, F. L., Watkins, J. C., Hammer, M. F., 'Global Genetic Variation at *OAS1* Provides Evidence of Archaic Admixture in Melanesian Populations', *Molecular Biology and Evolution*, 29, 6 (2012), pp. 1513–1520.

Mendez, F. L., Watkins, J. C., Hammer, M. F., 'Neandertal Origin of Genetic Variation at the Cluster of OAS Immunity Genes', *Molecular Biology and Evolution*, 30, 4 (2013), pp. 798–801.

Minster, R. L., Hawley, N. L., Su, C.-T., Sun, G., Kershaw, E. E., Cheng, H., Buhule, O. D., *et al.*, 'A Thrifty Variant in CREBRF Strongly Influences Body Mass Index in Samoans', *Nature Genetics*, 48, 9 (2016), pp. 1049–1054.

Montgomery, S. B., Dermitzakis, E. T., 'From Expression QTLs to Personalized Transcriptomics', *Nature Review Genetics*, 12, 4 (2011), pp. 277–282.

Narang, A., Jha, P., Rawat, V., Mukhopadhyay, A., Dash, D., Basu, A., Mukerji, M., 'Recent Admixture in an Indian Population of African Ancestry', *American Journal of Human Genetics*, 89, 1 (2011), pp. 111–120.

Nédélec, Y., Sanz, J., Baharian, G., Szpiech, Z. A., Pacis, A., Dumaine, A., Grenier, J.-C., *et al.*, 'Genetic Ancestry and Natural Selection Drive Population Differences in Immune Responses to Pathogens', *Cell*, 167, 3 (2016), pp. 657–669, e21.

Neel, J. V., 'Diabetes Mellitus: A "Thrifty" Genotype Rendered Detrimental by "Progress"?', *American Journal of Human Genetics*, 14 (1962), pp. 353–362.

Nica, A. C., Dermitzakis, E. T., 'Expression Quantitative Trait Loci: Present and Future', *Philosophical Transactions of the Royal Society of London B: Biological Sciences*, 368, 1620 (2013), 20120362.

Ongaro, L., Scliar, M. O., Flores, R., Raveane, A., Marnetto, D., Sarno, S., Gnecchi-Ruscone, G. A., *et al.*, 'The Genomic Impact of European Peopling of the Americas', *Current Biology*, 29, 23 (2019), pp. 3974–3986, e4.

Oota, H., Settheetham-Ishida, W., Tiwawech, D., Ishida, T., Stoneking, M., 'Human MtDNA and Y-Chromosome Variation Is Correlated with Matrilocal versus Patrilocal Residence', *Nature Genetics*, 29, 1 (2001), pp. 20–21.

Owers, K. A., Sjödin, P., Schlebusch, C. M., Skoglund, P., Soodyall, H., Jakobsson, M., 'Adaptation to Infectious Disease Exposure in Indigenous Southern African Populations', *Proceedings of the Royal Society B: Biological Sciences*, 284, 1852 (2017), 20170226.

Pagani, L., Kivisild, T., Tarekegn, A., Ekong, R., Plaster, C., Gallego Romero, I., Ayub, Q., *et al.*, 'Ethiopian Genetic Diversity Reveals Linguistic Stratification and Complex Influences on the Ethiopian Gene Pool', *American Journal of Human Genetics*, 91, 1 (2012), pp. 83–96.

Patin, E., Hasan, M., Bergstedt, J., Rouilly, V., Libri, V., Urrutia, A., Alanio, C., *et al.*, 'Natural Variation in the Parameters of Innate Immune Cells Is Preferentially Driven by Genetic Factors', *Nature Immunology*, 19, 3 (2018), pp. 302–314.

Patin, E., Lopez, M., Grollemund, R., Verdu, P., Harmant, C., Quach, H., Laval, G., *et al.*, 'Dispersals and Genetic Adaptation of Bantu-Speaking Populations in Africa and North America', *Science*, 356, 6337 (2017), pp. 543–546.

Patin, E., Siddle, K. J., Laval, G., Quach, H., Harmant, C., Becker, N., Froment, A., *et al.*, 'The Impact of Agricultural Emergence on the Genetic History of African Rainforest Hunter-Gatherers and Agriculturalists', *Nature Communications*, 5 (2014), p. 3163.

Patterson, N., Moorjani, P., Luo, Y., Mallick, S., Rohland, N., Zhan, Y., Genschoreck, T., Webster, T., Reich, D., 'Ancient Admixture in Human History', *Genetics*, 192, 3 (2012), pp. 1065–1093.

Petersen, D. C., Libiger, O., Tindall, E. A., Hardie, R. A., Hannick, L. I., Glashoff, R. H., Mukerji, M., *et al.*, 'Complex Patterns of Genomic Admixture within Southern Africa', *PLOS Genetics*, 9, 3 (2013), e1003309.

Piasecka, B., Duffy, D., Urrutia, A., Quach, H., Patin, E., Posseme, C., Bergstedt, J., *et al.*, 'Distinctive Roles of Age, Sex, and Genetics in Shaping Transcriptional Variation of Human Immune Responses to Microbial Challenges', *Proceedings of the National Academy of Sciences USA*, 115, 3 (2018), E488–E497.

Pickrell, J. K., Patterson, N., Barbieri, C., Berthold, F., Gerlach, L., Güldemann, T., Kure, B., *et al.*, 'The Genetic Prehistory of Southern Africa', *Nature Communications*, 3 (2012), p. 1143.

Pickrell, J. K., Pritchard, J. K., 'Inference of Population Splits and Mixtures from Genome-Wide Allele Frequency Data', *PLOS Genetics*, 8, 11 (2012), e1002967.

Pierron, D., Heiske, M., Razafindrazaka, H., Pereda-Loth, V., Sanchez, J., Alva, O., Arachiche, A., *et al.*, 'Strong Selection during the Last Millennium for African Ancestry in the Admixed Population of Madagascar', *Nature Communications*, 9, 1 (2018), p. 932.

Pierron, D., Heiske, M., Razafindrazaka, H., Rakoto, I., Rabetokotany, N., Ravololomanga, B., Rakotozafy, L. M., *et al.*, 'Genomic Landscape of

Human Diversity across Madagascar', *Proceedings of the National Academy of Sciences USA*, 114, 32 (2017), E6498–E6506.

Pierron, D., Razafindrazaka, H., Pagani, L., Ricaut, F. X., Antao, T., Capredon, M., Sambo, C., *et al.*, 'Genome-Wide Evidence of Austronesian-Bantu Admixture and Cultural Reversion in a Hunter-Gatherer Group of Madagascar', *Proceedings of the National Academy of Sciences USA*, 111, 3 (2014), pp. 936–941.

Pugach, I., Matveyev, R., Wollstein, A., Kayser, M., Stoneking, M., 'Dating the Age of Admixture via Wavelet Transform Analysis of Genome-Wide Data', *Genome Biology*, 12, 2 (2011), R19.

Qin, P., Stoneking, M., 'Denisovan Ancestry in East Eurasian and Native American Populations', *Molecular Biology and Evolution*, 32, 10 (2015), pp. 2665–2674.

Quach, H., Rotival, M., Pothlichet, J., Loh, Y. E., Dannemann, M., Zidane, N., Laval, G., *et al.*, 'Genetic Adaptation and Neandertal Admixture Shaped the Immune System of Human Populations', *Cell*, 167, 3 (2016), pp. 643–656, e17.

Quintana-Murci, L., Harmant, C., Quach, H., Balanovsky, O., Zaporozhchenko, V., Bormans, C., van Helden, P. D., Hoal, E. G., Behar, D. M., 'Strong Maternal Khoisan Contribution to the South African Coloured Population: A Case of Gender-Biased Admixture', *American Journal of Human Genetics*, 86, 4 (2010), pp. 611–620.

Racimo, F., Gokhman, D., Fumagalli, M., Ko, A., Hansen, T., Moltke, I., Albrechtsen, A., *et al.*, 'Archaic Adaptive Introgression in *TBX15/WARS2*', *Molecular Biology and Evolution*, 34, 3 (2017), pp. 509–524.

Racimo, F., Marnetto, D., Huerta-Sánchez, E., 'Signatures of Archaic Adaptive Introgression in Present-Day Human Populations', *Molecular Biology and Evolution*, 34, 2 (2017), pp. 296–317.

Racimo, F., Sankararaman, S., Nielsen, R., Huerta-Sánchez, E., 'Evidence for Archaic Adaptive Introgression in Humans', *Nature Review Genetics*, 16, 6 (2015), pp. 359–371.

Racimo, F., Sikora, M., Vander Linden, M., Schroeder, H., Lalueza-Fox, C., 'Beyond Broad Strokes: Sociocultural Insights from the Study of Ancient Genomes', *Nature Review Genetics*, 21, 6 (2020), pp. 355–366.

Raj, T., Kuchroo, M., Replogle, J. M., Raychaudhuri, S., Stranger, B. E., De Jager, P. L., 'Common Risk Alleles for Inflammatory Diseases Are Targets of Recent Positive Selection', *American Journal of Human Genetics*, 92, 4 (2013), pp. 517–529.

Ranciaro, A., Campbell, M. C., Hirbo, J. B., Ko, W. Y., Froment, A., Anagnostou, P., Kotze, M. J., *et al.*, 'Genetic Origins of Lactase Persistence and the Spread of Pastoralism in Africa', *American Journal of Human Genetics*, 94, 4 (2014), pp. 496–510.

Reich, D., Patterson, N., Kircher, M., Delfin, F., Nandineni, M. R., Pugach, I., Ko, A. M., *et al.*, 'Denisova Admixture and the First Modern Human Dispersals into Southeast Asia and Oceania', *American Journal of Human Genetics*, 89, 4 (2011), pp. 516–528.

Richerson, P. J., Boyd, R., Henrich, J., 'Colloquium Paper: Gene–Culture Coevolution in the Age of Genomics', *Proceedings of the National Academy of Sciences USA*, 107, Suppl. 2 (2010), pp. 8985–8992.

Rishishwar, L., Conley, A. B., Wigington, C. H., Wang, L., Valderrama-Aguirre, A., Jordan, I. K., 'Ancestry, Admixture and Fitness in Colombian Genomes', *Scientific Reports*, 5 (2015), p. 12376.

Robinson, M. R., Kleinman, A., Graff, M., Vinkhuyzen, A. E., Couper, D., Miller, M. B., Peyrot, W. J., *et al.*, 'Genetic Evidence of Assortative Mating in Humans', *Nature Human Behaviour*, 1, 1 (2017), p. 0016.

Rotival, M., Quach, H., Quintana-Murci, L., 'Defining the Genetic and Evolutionary Architecture of Alternative Splicing in Response to Infection', *Nature Communications*, 10, 1 (2019), p. 1671.

Rotival, M., Siddle, K. J., Silvert, M., Pothlichet, J., Quach, H., Quintana-Murci, L., 'Population Variation in Mirnas and Isomirs and Their Impact on Human Immunity to Infection', *Genome Biology*, 21, 1 (2020), p. 187.

Sams, A. J., Dumaine, A., Nédélec, Y., Yotova, V., Alfieri, C., Tanner, J. E., Messer, P. W., Barreiro, L. B., 'Adaptively Introgressed Neandertal Haplotype at the OAS Locus Functionally Impacts Innate Immune Responses in Humans', *Genome Biology*, 17, 1 (2016), p. 246.

Sankararaman, S., Mallick, S., Patterson, N., Reich, D., 'The Combined Landscape of Denisovan and Neanderthal Ancestry in Present-Day Humans', *Current Biology*, 26, 9 (2016), pp. 1241–1247.

Scepanovic, P., Alanio, C., Hammer, C., Hodel, F., Bergstedt, J., Patin, E., Thorball, C. W., *et al.*, 'Human Genetic Variants and Age Are the Strongest Predictors of Humoral Immune Responses to Common Pathogens and Vaccines', *Genome Medicine*, 10, 1 (2018), p. 59.

Schlebusch, C. M., Jakobsson, M., 'Tales of Human Migration, Admixture, and Selection in Africa', *Annual Review of Genomics and Human Genetics*, 19 (2018), pp. 405–428.

Schlebusch, C. M., Skoglund, P., Sjödin, P., Gattepaille, L. M., Hernandez, D., Jay, F., Li, S., *et al.*, 'Genomic Variation in Seven Khoisan Groups Reveals Adaptation and Complex African History', *Science*, 338, 6105 (2012), pp. 374–379.

Seielstad, M. T., Minch, E., Cavalli-Sforza, L. L., 'Genetic Evidence for a Higher Female Migration Rate in Humans', *Nature Genetics*, 20, 3 (1998), pp. 278–280.

Severe Covid-19, GWAS Group: Ellinghaus, D., Degenhardt, F., Bujanda, L., Buti, M., Albillos, A., Invernizzi, P., *et al.*, 'Genomewide Association Study of Severe Covid-19 with Respiratory Failure', *New England Journal of Medicine*, 383, 16 (2020), pp. 1522–1534.

Shah, A. M., Tamang, R., Moorjani, P., Rani, D. S., Govindaraj, P., Kulkarni, G., Bhattacharya, T., *et al.*, 'Indian Siddis: African Descendants with Indian Admixture', *American Journal of Human Genetics*, 89, 1 (2011), pp. 154–161.

Shelton, J. F., Shastri, A. J., Ye, C., Weldon, C. H., Filshtein-Sonmez, T., Coker, D., Symons, A., *et al.*, 'Trans-Ancestry Analysis Reveals Genetic and Nongenetic Associations with COVID-19 Susceptibility and Severity', *Nature Genetics*, 53, 6 (2021), pp. 801–808.

Siddle, K. J., Deschamps, M., Tailleux, L., Nédélec, Y., Pothlichet, J., Lugo-Villarino, G., Libri, V., *et al.*, 'A Genomic Portrait of the Genetic Architecture and Regulatory Impact of MicroRNA Expression in Response to Infection', *Genome Research*, 24, 5 (2014), pp. 850–859.

Sierra, B., Triska, P., Soares, P., Garcia, G., Perez, A. B., Aguirre, E., Oliveira, M., *et al.*, '*OSBPL10*, *RXRA* and Lipid Metabolism Confer African-Ancestry Protection against Dengue Haemorrhagic Fever in Admixed Cubans', *PLOS Pathogens*, 13, 2 (2017), e1006220.

Simonti, C. N., Vernot, B., Bastarache, L., Bottinger, E., Carrell, D. S., Chisholm, R. L., Crosslin, D. R., *et al.*, 'The Phenotypic Legacy of Admixture between Modern Humans and Neandertals', *Science*, 351, 6274 (2016), pp. 737–741.

Sironi, M., Clerici, M., 'The Hygiene Hypothesis: An Evolutionary Perspective', *Microbes and Infection*, 12, 6 (2010), pp. 421–427.

Skoglund, P., Malmström, H., Raghavan, M., Storå, J., Hall, P., Willerslev, E., Gilbert, M. T. P., Götherström, A., Jakobsson, M., 'Origins and Genetic Legacy of Neolithic Farmers and Hunter-Gatherers in Europe', *Science*, 336, 6080 (2012), pp. 466–469.

Skoglund, P., Mathieson, I., 'Ancient Genomics of Modern Humans: The First Decade', *Annual Review of Genomics and Human Genetics*, 19 (2018), pp. 381–404.

Skoglund, P., Thompson, J. C., Prendergast, M. E., Mittnik, A., Sirak, K., Hajdinjak, M., Salie, T., *et al.*, 'Reconstructing Prehistoric African Population Structure', *Cell*, 171, 1 (2017), pp. 59–71, e21.

Smith, M. W., O'Brien, S. J., 'Mapping by Admixture Linkage Disequilibrium: Advances, Limitations and Guidelines', *Nature Review Genetics*, 6, 8 (2005), pp. 623–632.

Snyder-Mackler, N., Sanz, J., Kohn, J. N., Brinkworth, J. F., Morrow, S., Shaver, A. O., Grenier, J.-C., *et al.*, 'Social Status Alters Immune Regulation and Response to Infection in Macaques', *Science*, 354, 6315 (2016), pp. 1041–1045.

Snyder-Mackler, N., Sanz, J., Kohn, J. N., Voyles, T., Pique-Regi, R., Wilson, M. E., Barreiro, L. B., Tung, J., 'Social Status Alters Chromatin Accessibility and the Gene Regulatory Response to Glucocorticoid Stimulation in Rhesus Macaques', *Proceedings of the National Academy of Sciences USA*, 116, (2019), pp. 1219–1228.

Souilmi, Y., Lauterbur, M. E., Tobler, R., Huber, C. D., Johar, A. S., Moradi, S. V., Johnston, W. A., *et al.*, 'An Ancient Viral Epidemic Involving Host Coronavirus Interacting Genes More than 20,000 Years Ago in East Asia', *Current Biology*, 31 (2021), pp. 3504–3514, e9.

Strachan, D. P., 'Hay Fever, Hygiene, and Household Size', *British Medical Journal*, 299, 6710 (1989), pp. 1259–1260.

Stranger, B. E., Montgomery, S. B., Dimas, A. S., Parts, L., Stegle, O., Ingle, C. E., Sekowska, M., *et al.*, 'Patterns of *Cis* Regulatory Variation in Diverse Human Populations', *PLOS Genetics*, 8, 4 (2012), e1002639.

Stranger, B. E., Nica, A. C., Forrest, M. S., Dimas, A., Bird, C. P., Beazley, C., Ingle, C. E., *et al.*, 'Population Genomics of Human Gene Expression', *Nature Genetics*, 39, 10 (2007), pp. 1217–1224.

Tang, H., Choudhry, S., Mei, R., Morgan, M., Rodriguez-Cintron, W., Burchard, E. G., Risch, N. J., 'Recent Genetic Selection in the Ancestral Admixture of Puerto Ricans', *American Journal of Human Genetics*, 81, 3 (2007), pp. 626–633.

Thomas, S., Rouilly, V., Patin, E., Alanio, C., Dubois, A., Delval, C., Marquier, L.-G., *et al.*, 'The *Milieu Intérieur* Study – An Integrative Approach for Study of Human Immunological Variance', *Clinical Immunology*, 157, 2 (2015), pp. 277–293.

Thornton, R., 'Aboriginal North American Population and Rates of Decline, ca. A.D. 1500–1900', *Current Anthropology*, 38 (1997), pp. 310–315.

Tung, J., Barreiro, L. B., Johnson, Z. P., Hansen, K. D., Michopoulos, V., Toufexis, D., Michelini, K., Wilson, M. E., Gilad, Y., 'Social Environment Is Associated with Gene Regulatory Variation in the Rhesus Macaque Immune System', *Proceedings of the National Academy of Sciences USA*, 109 (2012), pp. 6490–6495.

Vattathil, S., Akey, J. M., 'Small Amounts of Archaic Admixture Provide Big Insights into Human History', *Cell*, 163, 2 (2015), pp. 281–284.

Verdu, P., Becker, N. S., Froment, A., Georges, M., Grugni, V., Quintana-Murci, L., Hombert, J. M., *et al.*, 'Sociocultural Behavior, Sex-Biased Admixture, and Effective Population Sizes in Central African Pygmies and Non-Pygmies', *Molecular Biology and Evolution*, 30, 4 (2013), pp. 918–937.

Verdu, P., Pemberton, T. J., Laurent, R., Kemp, B. M., Gonzalez-Oliver, A., Gorodezky, C., Hughes, C. E., *et al.*, 'Patterns of Admixture and Population Structure in Native Populations of Northwest North America', *PLOS Genetics*, 10, 8 (2014), e1004530.

Vernot, B., Akey, J. M., 'Complex History of Admixture between Modern Humans and Neandertals', *American Journal of Human Genetics*, 96, 3 (2015), pp. 448–453.

Vernot, B., Tucci, S., Kelso, J., Schraiber, J. G., Wolf, A. B., Gittelman, R. M., Dannemann, M., *et al.*, 'Excavating Neandertal and Denisovan DNA from the Genomes of Melanesian Individuals', *Science*, 352, 6282 (2016), pp. 235–239.

Vicente, M., Priehodová, E., Diallo, I., Podgorná, E., Poloni, E. S., Černý, V., Schlebusch, C. M., 'Population History and Genetic Adaptation of the Fulani Nomads: Inferences from Genome-Wide Data and the Lactase Persistence Trait', *BMC Genomics*, 20, 1 (2019), p. 915.

Webster, T. H., Wilson Sayres, M. A., 'Genomic Signatures of Sex-Biased Demography: Progress and Prospects', *Current Opinion in Genetics & Development*, 41 (2016), pp. 62–71.

Wen, B., Li, H., Lu, D., Song, X., Zhang, F., He, Y., Li, F., *et al.*, 'Genetic Evidence Supports Demic Diffusion of Han Culture', *Nature*, 431, 7006 (2004), pp. 302–305.

Zeberg, H., Pääbo, S., 'The Major Genetic Risk Factor for Severe COVID-19 Is Inherited from Neanderthals', *Nature*, 587, 7835 (2020), pp. 610–612.

Zeberg, H., Pääbo, S., 'A Genomic Region Associated with Protection against Severe COVID-19 Is Inherited from Neandertals', *Proceedings of the National Academy of Sciences USA*, 118, 9 (2021), e2026309118.

Zhang, Q., Bastard, P., Liu, Z., Le Pen, J., Moncada-Velez, M., Chen, J., Ogishi, M., *et al.*, 'Inborn Errors of Type 1 IFN Immunity in Patients with Life-Threatening COVID-19', *Science*, 370, 6515 (2020), abd4570.

Zhernakova, A., Elbers, C. C., Ferwerda, B., Romanos, J., Trynka, G., Dubois, P. C., de Kovel, C. G., *et al.*, 'Evolutionary and Functional Analysis of Celiac Risk Loci Reveals Sh2b3 as a Protective Factor against Bacterial Infection', *American Journal of Human Genetics*, 86, 6 (2010), pp. 970–977.

Zhou, Q., Zhao, L., Guan, Y., 'Strong Selection at MHC in Mexicans since Admixture', *PLOS Genetics*, 12, 2 (2016), e1005847.

Epilogue

Adhikari, K., Chacón-Duque, J. C., Mendoza-Revilla, J., Fuentes-Guajardo, M., Ruiz-Linares, A., 'The Genetic Diversity of the Americas', *Annual Review of Genomics and Human Genetics*, 18 (2017), pp. 277–296.

Amariuta, T., Ishigaki, K., Sugishita, H., Ohta, T., Koido, M., Dey, K. K., Matsuda, K., *et al.*, 'Improving the Trans-Ancestry Portability of Polygenic Risk Scores by Prioritizing Variants in Predicted Cell-Type-Specific Regulatory Elements', *Nature Genetics*, 52, 12 (2020), pp. 1346–1354.

Boyle, E. A., Li, Y. I., Pritchard, J. K., 'An Expanded View of Complex Traits: From Polygenic to Omnigenic', *Cell*, 169, 7 (2017), pp. 1177–1186.

Bycroft, C., Freeman, C., Petkova, D., Band, G., Elliott, L. T., Sharp, K., Motyer, A., *et al.*, 'The UK Biobank Resource with Deep Phenotyping and Genomic Data', *Nature*, 562, 7726 (2018), pp. 203–209.

Cairns, J., *Matters of Life and Death*, Princeton, Princeton University Press, 1997.

Casanova, J. L., Abel, L., 'Inborn Errors of Immunity to Infection: The Rule Rather than the Exception', *Journal of Experimental Medicine*, 202, 2 (2005), pp. 197–201.

Fan, S., Hansen, M. E., Lo, Y., Tishkoff, S. A., 'Going Global by Adapting Local: A Review of Recent Human Adaptation', *Science*, 354, 6308 (2016), pp. 54–59.

Ge, D., Fellay, J., Thompson, A. J., Simon, J. S., Shianna, K. V., Urban, T. J., Heinzen, E. L., *et al.*, 'Genetic Variation in *IL28B* Predicts Hepatitis C Treatment-Induced Viral Clearance', *Nature*, 461, 7262 (2009), pp. 399–401.

Kerner, G., Patin, E., Quintana-Murci, L., 'New Insights into Human Immunity from Ancient Genomics', *Current Opinion in Immunology*, 72 (2021), pp. 116–125.

Krausz, C., Quintana-Murci, L., Rajpert-De Meyts, E., Jørgensen, N., Jobling, M. A., Rosser, Z. H., Skakkebaek, N. E., McElreavey, K., 'Identification of a Y Chromosome Haplogroup Associated with Reduced Sperm Counts', *Human Molecular Genetics*, 10, 18 (2001), pp. 1873–1877.

Martin, A. R., Gignoux, C. R., Walters, R. K., Wojcik, G. L., Neale, B. M., Gravel, S., Daly, M. J., Bustamante, C. D., Kenny, E. E., 'Human Demographic History Impacts Genetic Risk Prediction across Diverse Populations', *American Journal of Human Genetics*, 100, 4 (2017), pp. 635–649.

Martin, A. R., Kanai, M., Kamatani, Y., Okada, Y., Neale, B. M., Daly, M. J., 'Clinical Use of Current Polygenic Risk Scores May Exacerbate Health Disparities', *Nature Genetics*, 51, 4 (2019), pp. 584–591.

Mostafavi, H., Berisa, T., Day, F. R., Perry, J. R. B., Przeworski, M., Pickrell, J. K., 'Identifying Genetic Variants that Affect Viability in Large Cohorts', *PLOS Biology*, 15, 9 (2017), e2002458.

Ochi, H., Maekawa, T., Abe, H., Hayashida, Y., Nakano, R., Imamura, M., Hiraga, N., *et al.*, 'IL-28B Predicts Response to Chronic Hepatitis C Therapy – Fine-Mapping and Replication Study in Asian Populations', *Journal of General Virology*, 92, Pt 5 (2011), pp. 1071–1081.

Pizzol, D., Foresta, C., Garolla, A., Demurtas, J., Trott, M., Bertoldo, A., Smith, L., 'Pollutants and Sperm Quality: A Systematic Review and Meta-Analysis', *Environmental Science and Pollution Research International*, 28, 4 (2021), pp. 4095–40103.

Quintana-Murci, L., 'Human Immunology through the Lens of Evolutionary Genetics', *Cell*, 177, 1 (2019), pp. 184–199.

Sirugo, G., Williams, S. M., Tishkoff, S. A., 'The Missing Diversity in Human Genetic Studies', *Cell*, 177, 4 (2019), p. 1080.

Thomas, D. L., Thio, C. L., Martin, M. P., Qi, Y., Ge, D., O'Huigin, C., Kidd, J., *et al.*, 'Genetic Variation in *IL28B* and Spontaneous Clearance of Hepatitis C Virus', *Nature*, 461, 7265 (2009), pp. 798–801.

Thomas, S., Rouilly, V., Patin, E., Alanio, C., Dubois, A., Delval, C., Marquier, L.-G., *et al.*, 'The *Milieu Intérieur* Study – An Integrative Approach for Study of Human Immunological Variance', *Clinical Immunology*, 157, 2 (2015), pp. 277–293.

Torkamani, A., Wineinger, N. E., Topol, E. J., 'The Personal and Clinical Utility of Polygenic Risk Scores', *Nature Review Genetics*, 19, 9 (2018), pp. 581–590.

Zeberg, H., Pääbo, S., 'The Major Genetic Risk Factor for Severe COVID-19 Is Inherited from Neanderthals', *Nature*, 587, 7835 (2020), pp. 610–612.

Zeberg, H., Pääbo, S., 'A Genomic Region Associated with Protection against Severe COVID-19 Is Inherited from Neandertals', *Proceedings of the National Academy of Sciences USA*, 118, 9 (2021), e2026309118.

Index

Index

Index

gastroenteritis epidemics 143–4

Gauguin, Paul 1, 197

genes/genetics 3

  admixture and *see* admixture

  allele and 17–18, 19, 26, 41, 98, 102, 117, 139, 157–8, 160, 162, 191

  century of the 3

  clinical genetics 8, 145–6

  copy number variations *see* copy number variations

  disease/pathogens/immunity and *see* pathogens; immunity *and individual disease name*

  DNA and *see* DNA

  epigenetics 30, 181–6, 195

  gene, discovery of 13

  gene expression 24–5, 30, 163, 180, 181, 182, 186–8, 189, 195

  gene flow/migration 15, 20–21, 41, 80, 107

  gene loss ('pseudogenization') 140–47

  'genetic Adam and Eve' 5

  genetic diversity, sources of 16–21

  genetic drift 14, 15, 16, 17, 19–20, 98, 99, 100, 107, 109, 114, 119, 140, 168, 192

  genetic maps 65–6

  genetic sequencing 3, 4, 6, 10, 11, 22–5, 28, 34, 35–7, 38–9, 40, 41, 52, 53–4, 69, 70–71, 77, 78, 80, 85, 87, 99, 103, 109, 117, 120, 126–7, 138

  genetic transmission 12

  genetic variation *see* variation

  genetics, origins of 9, 15–17

  mutation *see* mutation

  natural selection and *see* natural selection

  origin of our species, first genetic data on 50–53

  population genetics *see* population genetics

  quantitative genetics 13–14

  SNPs *see* single nucleotide polymorphisms

  terms 13

  See also individual gene name

Genghis Khan 152, 171

genome/genomics 5, 6, 8, 25

  adaptation and *see* adaptation

  admixture and *see* admixture

  apes genome compared with human 27–30

  archaic human *see individual archaic human type*

  century of 3–4

  'coding' regions, genome 22–3, 24, 25, 40, 129, 134

  disease/pathogens/immunity and *see* pathogens; immunity *and individual disease name*

  diversity of genome 5, 9, 21, 91, 127, 203

  DNA and *see* DNA

  epigenome 182, 183, 185

  EWAS (epigenome-wide association studies) 183

  genetic diversity of a population, genomic factors of 17–19

  genome-wide association studies (GWAS) 127, 183, 187

  genomics defined 3

  human genome *see* human genome

  immunity and *see* immunity

  natural selection and *see* natural selection

  paleogenomic revolution 35–7, 56–7, 62, 64, 65, 69–72, 77, 78, 80, 81, 82, 85, 87

  pathogens and *see* pathogens *and individual disease name*

  revolution in genomics/Human Genome Project (2001) 3, 10, 22–3

  sequencing of genome 3, 4, 6, 10, 11, 22–5, 28, 34, 35–7, 38–9, 40, 41, 52, 53–4, 69, 70–71, 77, 78, 80, 85, 87, 99, 103, 109, 117, 120, 126–7, 138

  uniqueness of each individual and 23–6

  whole-genome studies 51–4, 58, 75, 102, 103, 110, 117, 120, 133, 146

Index

Index